ORGANIZATION STRUCTURE

CYBERNETIC SYSTEMS FOUNDATION

International Federation for Systems Research
International Series on Systems Science and Engineering

IFSR was established "to stimulate all activities associated with the scientific study of systems and to coordinate such activities at international level." The aim of this series is to stimulate publication of high-quality monographs and textbooks on various topics of systems science and engineering. This series complements the Federation's other publications.

A Continuation Order Plan is available for this series. A continuation order will bring delivery of each new volume immediately upon publication. Volumes are billed only upon actual shipment. For further information please contact the publisher.

Volumes 1–6 were published by Pergamon Press.

ORGANIZATION STRUCTURE

CYBERNETIC SYSTEMS FOUNDATION

Yasuhiko Takahara

Management and Information Sciences
Chiba Institute of Technology
Tokyo, Japan

and

Mihajlo Mesarovic

UNESCO Scientific Advisor on Global Change and
Case Western Reserve University
Cleveland, Ohio

Kluwer Academic / Plenum Publishers
New York, Boston, Dordrecht, London, Moscow

Library of Congress Cataloging-in-Publication Data

Takahara, Yasuhiko, 1935–
Organization structure: cybernetic systems foundation/Y. Takahara and M.D. Mesarovic.
p. cm. — (IFSR international series on systems science and engineering; v. 22)
Includes bibliographical references and index.
ISBN 0-306-47808-0
1. Industrial organization—Mathematical models. 2. Organization—Mathematical models. 3. Organizatonal sociology—Mathematical models. 4. System analysis. 5. Management science. 6. Cybernetics. I. Mesarovic, Mihajlo D. II. Title. III. Series.

HD30.25.T347 2003
302.3′5′0115—dc22

2003058898

ISBN 0-306-47808-0

233 Spring Street, New York, N.Y. 10013

http://www.wkap.nl/

10 9 8 7 6 5 4 3 2 1

A C.I.P. record for this book is available from the Library of Congress

Printed in the United States of America

To Late Professor Takehiko Matsuda

Contents

Preface

The objective of this book is to introduce a cybernetic system framework for the study of organizations using Mathematical General Systems Theory (MGST). The systems approach, i.e., a holistic perspective taking into account as many facets as possible yet within a "manageable" framework, has been used before in numerous studies of organizations such as [3] and [16].

The cybernetic approach has also been considered for the study of organization. In particular, as introduced in the cybernetic approach to management in *Organizational Transformation Learning* [6], they argued that cybernetics—the science of communication and control—can and should be used to present organizational problems in a common language. After all, an organization is formed for a purpose, to achieve a certain goal. Attempts to treat organizations mathematically have been reported in *A Mathematical Organization Theory* [16].

The use of mathematical general systems theory in this book is in sharp contrast to previous attempts to use the cybernetic systems approach and mathematics in organizational studies. The approach is distinct from numerical and computational methods used in management science in general where numerical branches of mathematics are exclusively used to solve specific problems. Our aim is to address "structural" problems of organization, i.e., the problems that are common to alternative classes/categories of organizations, to compare categories of organization in reference to their structural (class) properties and search for organizational improvement by structural change.

The book lays down a foundation for the initial efforts to investigate the potential of using **mathematical theory of general systems to address**

some of the well-recognized organizational dilemmas. Its full potential could be fully realized only by further investigation and research. It should also be pointed out that the benefits of the foundation introduced in this book are two-fold: It advances our understanding of organizations and also opens new avenues for research in MGST and Systems Science in general.

A project to develop a framework for the study of organization by cybernetics systems approach using the Mathematical Theory of General Systems, MTGS, started about fifteen years ago when the first author was asked to teach an undergraduate course entitled "Introduction of Management Engineering" at Tokyo Institute of Technology, Japan. Before teaching the course, the first author had been associated with a chair of management system, which belonged to the late Professor Takehiko Matsuda, who was a pioneer of scientific management and OR in Japan. This provided the motivation for the effort which lead to the publication of this book.

As indicated in the Appendix I of this book, there are at least three types of general systems theories. The authors of this work have been working on a formal theory, while the work of other authors on the application of the cybernetic concepts to organizations is primarily conceptual/philosophical. A natural question arises as to whether a formal theory of general systems can be applied to an organization or management. If yes, it would provide a theory of organization with a solid foundation. Conversely, since the application of a formal theory for that task is not as obvious, the application would inevitably enhance research in the formal theory of general systems as well. The project seemed promising.

At the beginning of the project, there had been several developed formal theories of organizations. In particular, Ref. [14] provides formal arguments for traditional theory like the scientific management; Ref. [16] develops a mathematical organization theory assuming that agents are input-output systems; Ref. [8] studies coordination theory where a goal seeking model is vital; and Ref. [17] deals with the adaptive behavior of a general goal seeking system in the view of the law of requisite variety. The most difficult part of the project was, then, to find a unifying framework to integrate those and other theories and to provide a comprehensive formal organization theory. Ref. [8] and Ref. [5] strongly influenced this book in this respect.

The project to develop a comprehensive formal theory of organization(s) is not an isolated attempt. There has been a movement called Computational and Mathematical Organization Theory (CMOT), aiming in developing such a comprehensive theory [1]. However, since the authors' project started independently of the other activities and since the authors' were motivated from different perspectives based on research interests and understanding of existing approaches to an organization theory, the results of this book are naturally different from those of CMOT. This book is developed using the formalized concepts of goal seeking system, decision principle, coordination principle, organizational culture and variety matching, and is totally based on the qualitative, non-numerical rather than numerical, mathematics.

It was claimed (c.f. [1]) that the objective of CMOT is to develop and test organizational theory using formal models. Although the objective of the project for this book is compatible with that of CMOT, the authors are more concerned with formalization of organizational concepts and development of a structural theory of an organization. The work reported in this book and the work of other approaches mentioned above could be considered as mutually complementary.

This book can be characterized as attempting to develop a new "grand theory". The results will ultimately depend on further, much more detailed studies. In particular, the book addresses several fundamental problems of current CMOT as addressed in Ref. [1]. The book provides a unified model and presents some insight into the relation between organizational structure and culture. It is the authors' view that if computational approaches are linked with the structural development in this book, a more powerful formal theory may emerge for the study of organizations. One of the authors' research groups is now engaged in studies in this direction [31,32]. Chap. 11 presents an example of such effort.

The contents of this book have been used for graduate courses at the universities (technically oriented) that the authors have been associated with. Although the book discusses only the most fundamental topics of an organization, the material developed so far and reported in this book is ample enough to be taught as a formal organization theory course in other universities.

The authors owe development of this book to many of their friends and colleagues who help them with the project. In particular, Prof. Bunpei Nakano, Tokyo Institute of Technology, Japan, must be highly appreciated. Many results of this book come from the works he kindly introduced to the first author. Without his help the project could not have even started. Prof. Xiaohong Chen of Central South University, China, must be also deeply thanked, who co-worked with the first author and gave him many opportunities to lecture the contents of this book at her graduate school.

At the final stage of preparation of the draft, Prof. Franz Pichler of University of Linz, Austria, provided the first author with the most important opportunity to revise the contents thoroughly. Due to his kindness, the first author delivered lectures using the material of the original draft. The title of the book was planned to be "A Mathematical Organization Theory", it was finalized as it is now through the revising work at Linz.

The authors must last but not least thank Dr. Caroline Benton, Tsuchiya Planning Company, Japan and Mr. Yoshio Yano, Ph.D. candidate, Chiba Institute of Technology, Japan. Dr. Caroline Benton checked the earliest version of the draft prepared by the first author thoroughly with her professional knowledge in management theory so that it can be understandable. The current book is certainly an offspring of that version. But for expert knowledge in Microsoft Word of Mr. Yano the project could not have been finished.

Spring, 2003

Authors

CHAPTER 1 ORGANIZATIONAL CYBERNETICS

The objective of this book is to introduce a cybernetic framework for the study of organization using Mathematical General Systems Theory, MGST.

An organization is formed for a purpose, to achieve a certain goal by making decisions pursuant of that goal. It is therefore also a natural candidate for a cybernetic perspective where cybernetics as introduced by Norbert Wiener is a study of "information and control" decision making systems in "man and machine". The cybernetic approach has also been the subject of the study of organizations, e.g., [5] and [6] but not in a formal, mathematical, manner.

The systems approach, i.e., a holistic perspective taking into account as many facets as possible yet within a "manageable" framework, has been used before in numerous studies of organizations such as [3] and [16]. The use of mathematical general systems theory is in sharp contrast to previous attempts to use the systems approach in organizational studies. They are one of either of two categories: either conceptual, mostly verbal, recognizing uncertainty and human factors, or they consist of the application of numerical and computational methods to address specific organizational problems.

Our concern is to address "structural" problems of organization, i.e., the problems that are common to alternative classes/categories of organizations, to compare categories of organization in reference to their structural (class) properties and search for organizational improvement by structural change. Non-numerical branches of mathematics are used and the analysis proceeds in a top-down manner. Starting from a general classification of organization, additional mathematical structure is added to develop a deeper understanding.

The book lays down a foundation for the initial efforts to investigate the potential of using mathematical theory of general systems to address some of the well-recognized organizational dilemmas. Its full potential could be fully developed only by further investigation and research. It should also be pointed out that the benefits of the foundation introduced in this book are two-fold: It advances our understanding of organizations and also opens new avenues for research in MGST and Systems Science in general.

MGST has been applied mainly for addressing engineering systems [2]. Since these systems are well structured compared to social systems, they were easier candidates for the application of a formal — mathematical — theory. An organization is, on the contrary, ill structured and hence provides a real challenge to MGST.

In this chapter first we introduce two earlier efforts to apply the systems approach to the study of organizations in order to compare with and relate to the approach in this book. One is a general approach while the other uses cybernetic concepts. Neither of these approaches is mathematical. They are representative of how the systems approach in the study of organizations has been used so far.

Next we discuss the concept of a multi-layer system in MGST. Because of the apparent similarity between standard models of organization theory and multi-layer systems of MGST, the latter could be misunderstood as having been developed as a formalization of an organization. Actually, the MGST model was developed independently. The similarity, however, suggests that the development of a formal theory of an organization based on MGST is possible.

Finally, we present a basic cybernetic model of an organization based on MGST. The model is particularized to conform with the perspectives of the two traditional systems approaches to the study of organization presented here.

Appendix I presents a brief introduction to MGST to make the book self-contained [2].

1.1 Traditional Systems View of Organization

An organization as presented in Ref. [3] is characterized as:

(i) An open socio-technical system;
(ii) A contrived system;
(iii) A hierarchical system;
(iv) Having negative entropy;
(v) Being either in steady state or at dynamical equilibrium;
(vi) Having a feedback mechanism;
(vii) Adaptive and maintenance mechanism;
(viii) Growth through internal elaboration;
(ix) Equifinality of an open system;
(x) Managerial system;

An organization is viewed as an open system that interacts with environments. It is affected by the environment and conversely affects the environment while maintaining its own proper functioning.

The external behavior of an organization is modeled as an input/output system. We recognize two types of inputs: The first type is a resource input such as personnel, material, money, energy or information. The second is external managerial information related to customer demands, consumer behaviors, marketing conditions, economic situations, etc. This classification is crucial for our approach. This classification of inputs into a resource input and information input is not a novelty. The best example is the living systems theory [4]. An organization transforms the resource inputs into products or services and delivers the outcome to the environment. The transformation, which usually requires support of a specific technology, is a primary function of an organization.

In this book — with the exception of the garbage can model given in Chap. 11 — it is assumed that the resource input is a controllable input and hence it can be treated as one component of the internal decision variable set or as a constraint to the management decision process. It will not be treated as a real external input to an organization because an external input in the systems theory is uncontrollable. Since the external information input is assumed uncontrollable, it will be treated as the only real input to the organization. In Chap. 11, however, the constraint is treated as an additional external input.

An organization is a stable system. If it were unstable, it would have become extinct. There are two types of stability: behavior stability and structural stability. The behavioral stability is sustained by a feedback mechanism. On the other hand, the structural stability—meaning to sustain characteristic parameters of the organization — is realized by higher-level management activities mentioned below.

The importance of negative entropy has been claimed to contribute to the structural stability. A typical example of a negative entropy is a newly recruited employee. Without the import of negative entropy, it is asserted that an organization cannot avoid thermal death. The stability provides a steady state behavior of an organization.

An organization is a contrived system. It is an artificial system designed purposefully to realize a pre-assigned goal, which will be referred to as an organizational goal. The existence of an organizational goal is a fundamental characteristic of an organization.

An organization is a managerial system whose management activity is to control the primary function of transformation of inputs to outputs while pursuing the organizational goal. Management is a human task while the transformation is performed with help of technology tools. An organization is essentially a man-machine system.

Control of the transformation requires collaboration of more than two persons (specialists). Higher-level management is then needed to coordinate that collaboration. The management of an organization has to have a hierarchical structure in order to accomplish that. An organization is a socio-technical system more than a simple man-machine system.

Coordination management is also required by the fact that each member of an organization is assumed to have its own goal. This is another fundamental assumption about an organization. A member of an organization is not a robot. Therefore, there can be conflicts among members as well as between the organizational goal and the goal of its members'. The resolution of these conflicts is a basic problem in organization theory.

An organization must have an ability to survive and grow in a changing and hostile environment and properly address the managerial information input. This ability is called adaptation ability or self-organization ability. Then, there must be a layer in the management hierarchy that is responsible for adaptation or self organization.

Conventionally, then, three hierarchical layers have been recognized in organization theory: Technical level, organizational level and institutional level. According to Ref. [3] the three layers are structured like a shell as shown in Fig. 1.1.

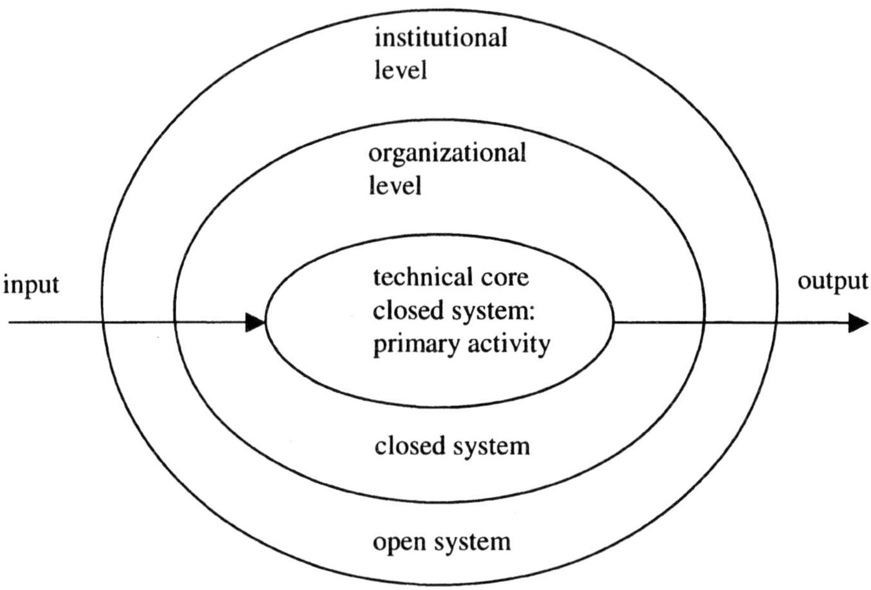

Figure1.1 Three layers of organization

The technical level, which is the innermost layer, is protected from the environment by the other levels and is concerned with the primary operational task of an organization, i.e., the transformation of the resource input into organization's product. Its activity is characterized by economic/technical rationality forming a closed technical core. Since the uncertainty of the environment is irrelevant to this level, a closed system view is applicable.

The institutional level, which is the outermost layer, faces the environment — the information input — and absorbs the uncertainty that is a characteristic of the openness of an organization. An open system view must be applied to the institutional level responsible for adapting to the environmental change by modifying organizational parameters and structure. This level constitutes the adaptive and maintenance mechanism of an organization. The organizational goal is transformed by this level into an operational goal, called the global goal, depending on the state of the environment.

The organizational level is a mediator between the institutional level and the technical level. The global goal provides a guideline for the activity of the organizational level. The organizational level coordinates behaviors of the technical core by controlling allocation of resources to the technical level.

Each level has managers. The roles of managers are characterized according to functions of the levels.

Managers of the technical level are specific task oriented with a short time perspective. They are concerned with the efficiency of the transformation (production) evaluated by economic/technical rationality given available technology.

Managers of the organizational level are, as mentioned above, mediators between the other levels. Their activity is to find an optimal compromise to harmonize lower level activities and the upper level demand represented by the global goal.

Managers of the institutional level should have a broad conceptual frame of reference. They control organizational behavior and structure, adapting them to the environmental change. Their activity is judgmental due to the uncertainty of the environment.

1.2 Management Cybernetics: A Viable System Model

The management cybernetics is an application of the cybernetics, (i.e., decision making concepts) to management providing more specific representation of the management functions [5,6]. The viable system model, VSM, of Ref. [5] shown in Fig. 1.2 is an example of such a representation.

The VSM is a hierarchical system consisting of five layers. It is an open system that interacts with its environment. The first layer carries out the basic task of an organization, i.e., the transformation of resources. It is assumed that the task of the first layer is departmentalized and the layer consists of local operational units. Each local operational unit is composed

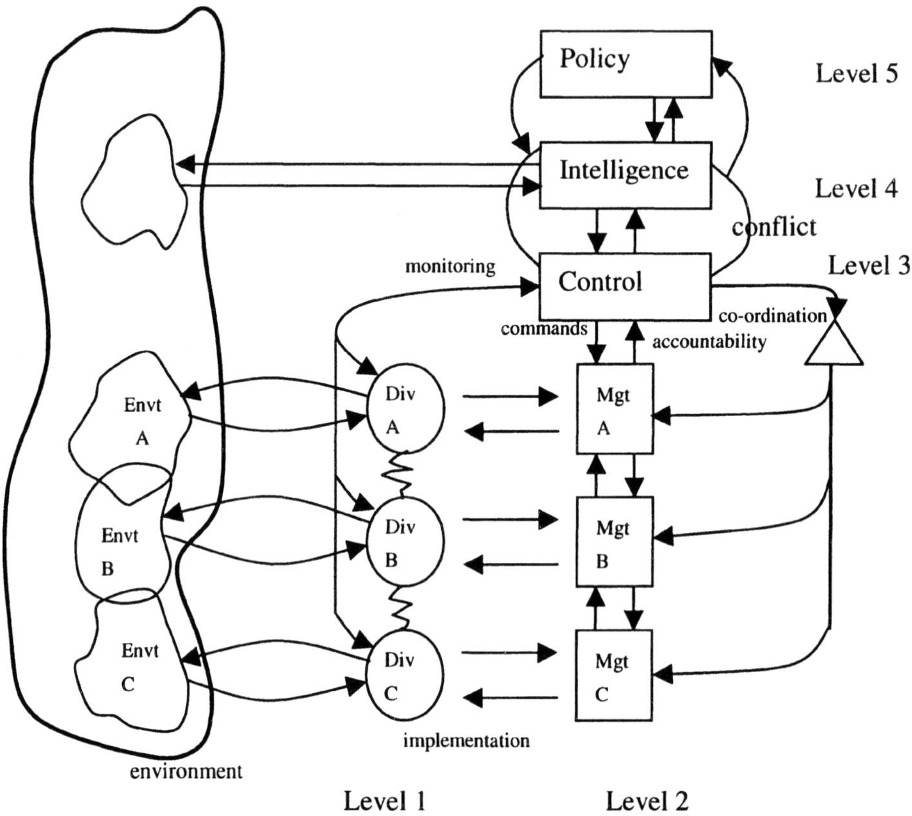

Figure1.2 Viable system model (Ref. [5])

of two components: a local process and its operator. Local processes interact with each other constituting the total process of transformation and are assumed open to the environment.

In VSM the regulatory activities (feedback control) are emphasized while at the same time the local operational units are allowed to have autonomous behaviors. Therefore, conflicts naturally exist among them. A coordination scheme is needed to resolve the conflicts. The scheme produces a coordination plan after negotiation process between the first layer and the third layer mentioned below. Each local operational unit is associated with one local system unit which implements a given coordination plan through a local feedback mechanism.

Since the associated coordination task unit of each local operational unit has to negotiate with other associated units to execute the coordination plan, it can be considered to be an interface of the local operational units. The combination of a local operational unit and its associated process will be called in this book a local operational level management unit.

The third layer is responsible for resolution of internal conflicts and for solution (decision making) of short range problems. It must produce a coordination plan negotiating with the first layer. Coordination is realized by management resources allocation.

The second layer or system executes the produced coordination plan under supervision of the third layer. The second layer actually implements the plan faithfully co-working with and controlling the associated units of the first layer. In this sense VSM defines the function of the second layer as a regulatory and anti-oscillatory one.

At the execution stage of the coordination plan the third layer controls the first level units by auditing their activities assuring their accountability.

The important feature of the third layer is that it is assumed to be isolated from the uncertainty of the environment, just like the first layer. The third layer is mainly concerned with the internal and current state of the organization. The third layer is separated from 'spatial' uncertainty (the environmental uncertainty) and uncertainty about the future.

The fourth layer or system takes care of the long-term uncertainties of the environment—spatial and temporal—and produces a strategy for organizational adaptation. The key function of this layer is self-organization.

The fifth layer or system is the brain of the organization, which resolves the conflict between the third layer and the fourth layer, and produces a policy for the organization. The third layer is concerned with the current efficient operation of the primary activity of an organization, while the fourth layer is concerned with adaptation for the future activities. Although both are important, there can be conflict between them.

The hierarchical structure constitutes a chain of meta-functions. It is assumed that the chain is terminated by the fifth layer.

The VSM can be represented as a system consisting of three functional layers, operation (process), operational control and managerial control. The managerial layer covers the third to fifth layers while the operational layer form the first to the third levels. The third layer lies in both layers as a mediator.

The VSM emphasizes that this three layer structure is nested recursively in a viable system. Fig. 1.2 indicates that we can identify a three layer structure in the operation and the operational control units which is homomorphic to the original hierarchy. This recursion displays in the “Chinese box” structure of MGST.

Every organization theorist admits that an adaptation concept is indispensable to organization theory. But in many cases we do not have a definite statement about what the adaptive function really means. In the VSM, however, adaptation is particularized by the variety concept. Ref. [7] proposed the law of "requisite variety" as a system theoretic law, which insists that a system must have at least the same degree of behavioral variety as its environment does in order to survive. That is, the number of behaviors a system can exhibit must not be less than that of the environment.

The VSM uses the requisite variety concept as the fundamental principle to design and to explain management activities. It claims that a viable system should satisfy the following principle [5]:

"Managerial, operational and environmental variety, diffusing through an institutional system, tend to equate; they should be designed to do so with minimum damage to people and to cost".

Ref. [5] asserts that a surviving organization has functions to decrease the variety of environmental input by some “market research” mechanism, to reduce information input according to organizational objectives and to increase the variety of organizational behavior by appropriate “market” means. The former function is called variety attenuation while latter function is considered as variety amplification. In this way an organization can adapt to environmental change.

When the VSM is compared to the systems approach of Sec. 1.1, the local units correspond to the technical level, the third layer to the organizational level, and the other layer to the institutional level. The VSM presents a detailed image of the management function of the management hierarchy.

1.3 General Systems Theoretic View of Complex System: Multi-layer Model

The concept of a multi-layer system in GST as an exemplification of a complex system, shown in Fig. 1.3 [8] is the starting point for the cybernetic theory of organization developed in this book.

The hierarchy has four layers: Process, regulation/optimization, coordination, and adaptation/self-organization. These layers are identified by their control functions.

The process layer corresponds to the operation layer of the VSM, and is supposed to consist of sub-processes that are modeled as input output systems. They usually interact with each other.

The regulation/optimization layer corresponds to operational control. This layer is responsible for control of the process layer and is concerned with optimum efficient operation on an economic and rational basis. This layer is also decomposed into subunits. The subunits are called decision makers or controllers. The regulation function is usually implemented by a

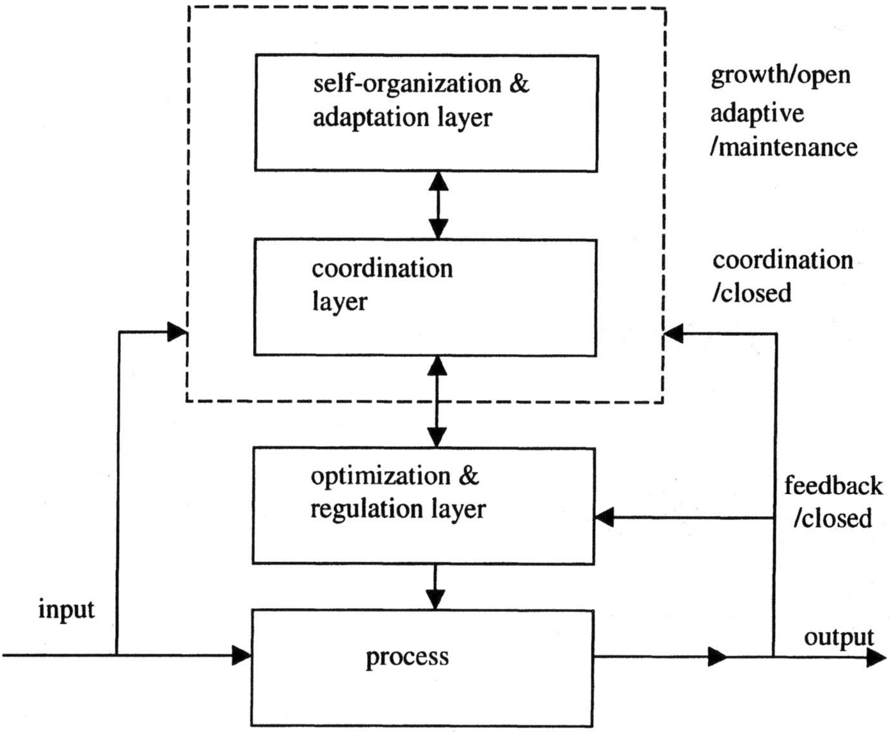

Figure1.3 Multi-layer model of MGST

feedback mechanism. The regulation/optimization layer relies on the closed system of its operation.

Since interactions exist among the sub-processes, the regulation/optimization layer cannot perform its function without proper information regarding interactions. Furthermore, if the regulation/optimization layer is given autonomy, coordination is necessary for conflict resolution among the controllers. The coordination function of the third layer (coordination layer) provides the second (optimization/regulation) layer with information regarding interaction, and integrates the behaviors of the second layer to realize the goal specified by the upper level.

The adaptation/self-organization layer, the fourth layer, absorbs the uncertainty of the external input and transforms it into a well-defined task —

without uncertainty — which is transmitted to the third layer as a system parameter. For instance, an operationalized goal of the organizational goal is an example of a system parameter.

If we notice that the combination of the coordination level and the adaptation/self-organization level corresponds to the management layer of the three layer model of VSM, we can see that the MGST hierarchical structure becomes a restatement of the VSM.

Various formal theories have been developed for the hierarchical layer model of MGST mostly applied to engineering systems. The regulation function is a principal topic of the control theory. The optimization function has been extensively studied in the control theory and the management science [31]. For the coordination function formal results have been derived in the multi-level systems theory [8]. The study of the adaptive function has a long history in the control engineering, where system parameter adjustment is a main topic of adaptive control theory.

Although these formal theories have rich results, they cannot be said to satisfactorily address problems of an organization because they are mostly developed for well-structured systems in a quantitative way, while an organization system is typically ill-structured at least quantitatively. This book will mainly pursue qualitative development different from the conventional formal theories.

1.4 Model of Organizational Cybernetics

In this book we develop a formal theory of organization as an application of MGST. The results are more conceptually oriented or more qualitative than those of conventional control theory or management science.

The starting point of the development is the hypothesis that if general concepts of MGST are properly specialized, useful mathematical theory based on the results in MGST can be developed. We will try to demonstrate that this hypothesis is true and will be even more so in the future.

Fig. 1.4 shows the basic model we use for the development of organizational cybernetics. The model is derived from the MGST complex specialized to meet the characteristics of an organization.

The organizational cybernetics model characterizes an organization as follows:

(i) From the external perspective, an organization is an input/output system open to its environment. As mentioned earlier, there are two types of inputs from the environment: Resource inputs and managerial information inputs. The resource input is assumed controllable, and hence is not a real input in the sense that there is no uncertainty: Although it is coming from outside, the inflow is determined by the organization. The resource input is a component of decision variables or a constraint to decision making. The managerial information input is assumed uncontrollable. If an uncontrollable

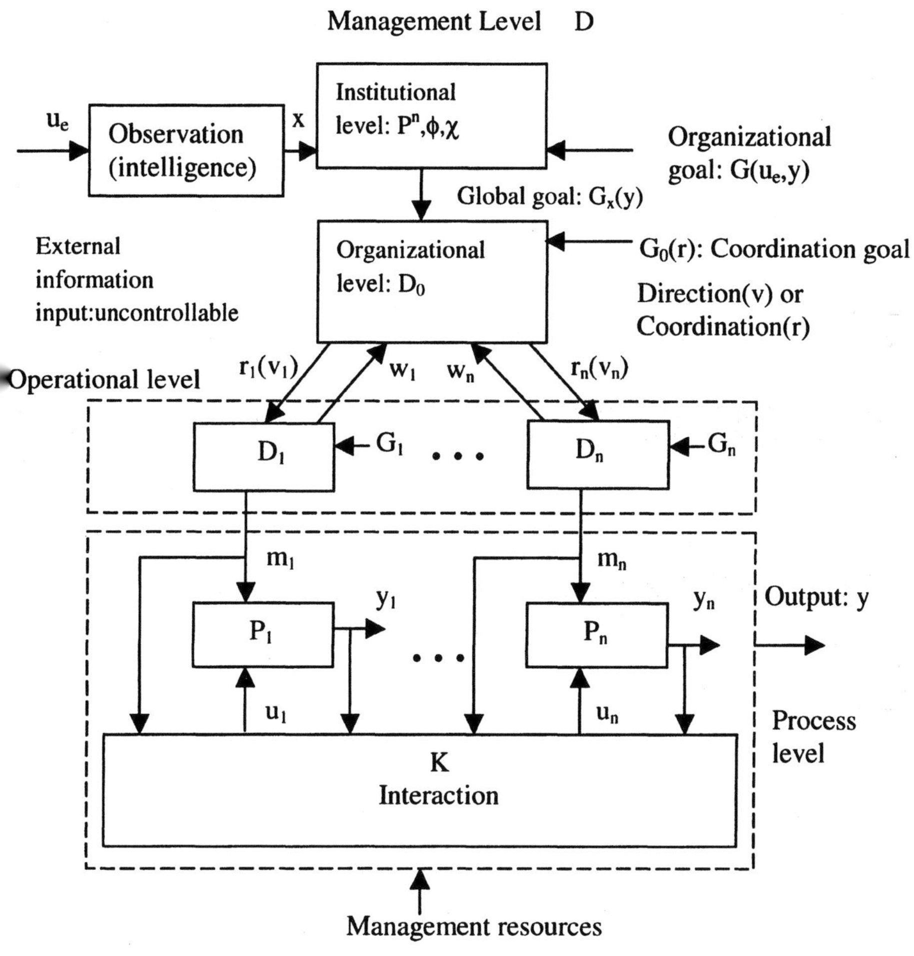

Figure1.4 Basic model of organization

aspect exists for the resource input, it is included into the managerial information input as one of its components. A real input — possessing uncertainty — to the organization is, hence, the managerial information input.

The managerial input will be further divided into two types: environmental process information and environmental structure information. They are observed by respective observation mechanisms that can exist as real objects or may be in the mind of a decision maker.

The information input resulting from the observation process is used for tactical adaptation, while that of the structural information input is used for strategic adaptation. The observations are subject to organizational learning.

The output of an organization will be called a product.

(ii) An organization is a dynamic system. Variables that describe an organization are assumed to be time functions defined in Chap. 5.

(iii) Internally, the model consists of four levels: process level, operational level, organizational level and institutional level.

(iv) The resource input is transformed into a product by the process level subsystem as a primary activity of an organization.

(v) The process activities are departmentalized. It is, therefore, composed of sub-processes. The sub-processes interact through the interaction block K.

(vi) The operational level controls the process and is responsible for its rational (economical) behavior, and hence is also departmentalized into subunits according to the departmentalization of the process level. Each subunit will be called an operational level decision maker that has its own goal and executes an optimization function to pursue the goal. In this sense it has an autonomy.

It is important to notice that the goals of the operational level are in general distinct from the organizational goal. The operational level is responsible for stable operation of the process. The stability is achieved by a regulation (feedback) function.

(vii) An organization has a goal called organizational goal. The organizational goal is a global evaluation of the transformation activity of the process.

(viii) The institutional level addresses the managerial information input. The input is modeled as consisting of two components: Structural change and steady state noise. The institutional level is mainly concerned with the structural change. It is assumed to have a normative image about the operational model P^m. An adaptation strategy is based on that model so that the organizational goal can be transformed into an operational goal, which will be called a global goal.

(ix) The organizational level is concerned with coordination of the operational level. In general, successful coordination does not imply realization of the global goal. If successful coordination implies realization of the global goal, the organizational structure is said to be valid. One of the main topics of organizational cybernetics is to find conditions for the design of a valid organizational structure.

(x) A harmonious relation between the global goal and the goals of the operational level will be formalized as goal compatibility. It will be shown that compatibility is essential for a valid organizational structure.

(xi) The validity condition provides a link between the micro and macro problems of organization theory, where the optimization activity of the operational level corresponds to the micro level issues, while organization behavior controlled by the global goal corresponding to the macro level. The organizational image of this book presents a comprehensive view of integrated behavior of the micro and the macro levels.

Chap. 11 will discuss the "garbage can model" of an organization which is considered drastically different from a rational model which is the main subject of this book [10]. Essentially, the garbage can model does not have a hierarchical structure analogous to the organizational cybernetic model. The model has two inputs: choice opportunity and problem. Both inputs change dynamically and at random. In Chap. 11 the global goal is interpreted as the choice opportunity and the constraint is interpreted as the problem. The garbage can model is treated as a three level structure without a management level. In this way, the garbage can model will be shown to be embedded into the organizational cybernetic model, where decision making and problem solving are done according to how problems and operational units are attached to a choice opportunity.

In Sec. 1.1, we introduce key terms of the traditional current literature on the systems approach to organization quoted earlier [3]. It will be shown that the model in Fig. 1.4 will encompass the basic concepts of the traditional systems approach to organization: Contrived system, hierarchical system, feedback mechanism, adaptive and maintenance (coordination) mechanism and managerial system.

The most significant aspect that is missing in the model of this book is investigation of growth through internal elaboration. However, in order to include the growth aspect in our mathematical theory of organizational cybernetics, we need a more solid theory of self-organization. Although many researchers have discussed self-organization it cannot be said that there is a satisfactory formal theory suitable for our purpose. It is doubtful that the concepts of negative entropy and equifinality of open systems can play a vital role in a formal theory of organization, although they can be useful at times as a persuasive tool in verbal discussions about an organization.

CHAPTER 2 GOAL SEEKING SYSTEM

The essential character of an organization as well as its members lies in their goal seeking activities. A goal seeking activity covers a wide range of behaviors, for instance, from a rational activity considered in this chapter to the activity of the garbage can model of Chapter 11. Therefore, it is not easy to describe the activity in simple terms. However, a unified formal model for it is a prerequisite to develop a formal theory of the cybernetic approach to an organization. Using GST concepts presented in Appendix I a formal description of the goal seeking activities in an organization is introduced in this chapter providing a general conceptual model for an organization.

We first examine a model of an organization member's behavior. Since members constitute basic components of an organization, this examination is important for the organization theory. At the same time it will be shown that the behavioral model provides a concrete image of the formal GST model of Sec. 2.3, which, in turn, is an abstract mathematical object. The illustration will show how the formal model is related with a real-life organization providing its meaningful representation.

Next, a goal seeking system of GST will be presented as a formal model of the goal seeking activity. A goal seeking system consists of two components: A causal system representing a process, procedure, activity, etc., and a decision maker. The process is the target for which a decision is to be made. It is modeled as an input/output system whose input is composed of two components: An input from its environment and an input from the decision maker. The former input—which will be called the external input—is not supposed to be under the control of the decision maker while the latter input—which will be called the decision variable—is completely managed by the decision maker. The external input represents an uncertain fact or the ill-structuredness of the goal seeking activity while the decision variable may be considered as a parameter of the process to be adjusted to address the uncertainty. The output of the process represents the outcome of the goal seeking activity.

The decision maker is characterized by a decision model and a decision principle. The decision model consists of an image of the process and an evaluation function of the behavior of the process. The process model—which can be different from the real process—may or may not contain an uncertainty factor.

A decision principle is an expression of a decision making or a problem solving philosophy to deal with the ill-structuredness. It makes it possible for the decision maker to transform a model with uncertainty into one without uncertainty and makes it possible to produce a logical decision.

Finally we will investigate the relation between the internal structure and the external behavior of a goal seeking system. The external behavior corresponds to an input output description of an organization or that of its member. This description represents a standard operating procedure of the bureaucratic model [18] and will be used in the discussion of a normative bureaucratic view of an organization in Chapter 9 and Chapter 10.

2.1 Organization Member as Goal Seeking System

Fig. 2.1 shows a systems approach based perception of the behavior of an organization member where behavior is understood to mean how the system (member) responds (behaves in response to) to the "prevailing conditions", i.e., external stimuli [3].

Although the behavior is dependent on the stimulus this dependence is not straightforward.

First, it is "filtered" through the processes that include: perception, cognition and motivation.

Second, the behavior is not solely determined by the member itself but is conditioned by circumstances and past experience.

An organization member does not accept a stimulus as it is but through its perception facility which gives a meaning to the stimulus. What really affects the behavior of a member is a perceived stimulus rather than the stimulus itself. The perceived stimulus is processed by the "cognitive" process; i.e., knowledge/skill of the member resulting in (producing corresponding) appropriate behavior. The cognitive process that will be elaborated on later is based on an *image* the member has of the world external to it and the *selection* of a behavior appropriate to the circumstances.

Usually there are more than one alternative for the selection of the behavior.

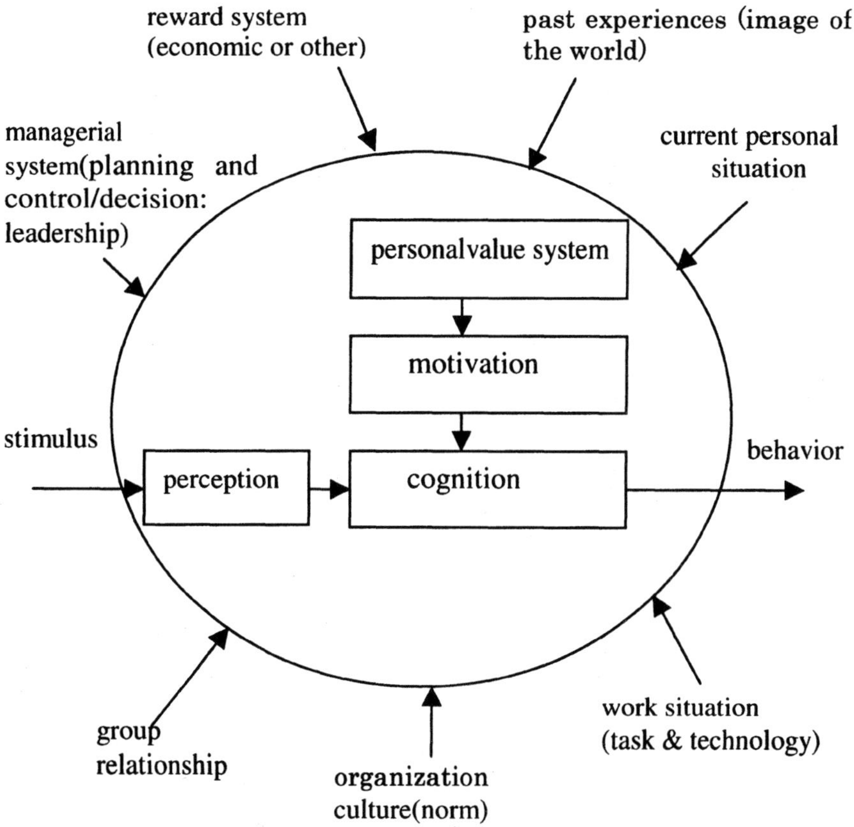

Figure 2.1 Model of organization member

Third, three aspects—perception, cognition and motivation—are not isolated. They are influenced by environmental conditions and what the member has learned from the past. Principal among them are indicated in Figure 2.1: Work situation, reward system, managerial system, group relationship, organization culture, current personal situation and the past experiences of the member, etc.

The working situation includes physical and psychological working conditions, formal organizational relationships, technological situations and others. The managerial system determines decision and leadership styles of members.

Since the perception, cognition and motivation of organization members depend on propensities specific to individuals, the behaviors of different members cannot be the same even if they receive the same stimulus. This, in particular, is true for motivation that is a key factor in explaining the differences in behavior.

Typically, the behavior of an organization member is characterized as being [3]:

- Causal
- Goal-directed
- Motivated

Although organization members behave in diversified ways, a common framework can be conceived on an abstract level.

The mathematical model of an organization member of this book will be built on this abstract level.

Causal characterization refers to the response of the member to stimuli. The overall behavior of the member essentially is a causal input/output system.

The perception is represented by an observation process that interprets the external input. The cognition is modeled by a parameterised input/output system called the image; it is a model the decision maker has of the causal process and its responsibility. The parameter will be called decision variable. The motivation is represented by an assessment, evaluation, function which provides a criterion for the selection of the parameters. It is important to recognize the basic indeterminancy of the member's behavior which is not the same if the same stimulus is presented in two different instances. The input/output representation is not a determined system. This is due to the system (members functioning) as will become clearer subsequently. The input/output—stimulus/behavior—transformation depends on the state of the system at the time, on a deeper structure, i.e., on the state of perception, cognitive and motivational processes. The motivation, in turn, depends on the goal, which is also not invariant over time and could reflect past experience from the evaluation of the system in time. The personal value of the member is represented by a decision principle that specifies which of the alternative sets of criteria is to provide the motivation. Recognition of a decision principle as a personal choice of an organization member is generally not well recognized.

These concepts constitute the framework of a system theoretic model of an organization member.

In reference to the circumstances in which a member of organization finds itself and its attitude, an organization can be characterized in two ways:

In Theory-X, members of an organization are not responsive or are even resistant, antagonistic, to the organizational needs. They must therefore be persuaded, rewarded, punished, controlled, etc., so that their activities are directed to fit the needs of the organization.

In Theory-Y the members are not passive or resistant per se. They become negative as a result of experiences in the organization. Therefore, the essential task of management is to arrange organizational conditions and methods of operation so that the members can achieve their goals best while directing their own efforts toward the organizational objective as well.

These views are naturally related with the managerial style of an organization.

This book will explore management based on Theory-Y. If an organization is valid in the sense of Chapter 1, organization members are allowed to make the best use of their talents to realize their own goals while at the same time their optimum performances lead to an optimum of the organizational goal also. The book is, therefore, particularly interested in how an organization can be made valid in the sense outlined in Chapter 1.

2.2 Decision Problem

A fundamental system theoretic model to explain the behavior of an organization member is the goal seeking model of GST. The starting point for the representation of the structure of a goal seeking model is shown in Fig. 2.2.

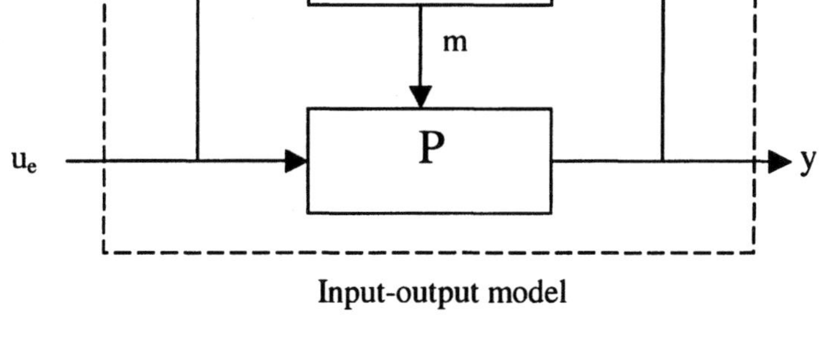

Figure 2.2 Goal seeking system

A goal seeking model consists of two basic parts: Causal subsystem P and Decision making subsystem D. The former will be referred to as the process while the latter will be referred to as the decision maker. The process P is an object that is a target of the decision-making activity of D.

D and P can be interpreted in several ways. One interpretation given in Sec.2.1 P corresponds to an issue for decision making while D corresponds to an organization member itself who is characterized by perception, cognition model, motivation and personal values. The cognitive model, which is an image of P, will be denoted by P^m. Ideally, P^m should be P repressed as close as possible although that is never the case in reality either because of inherent uncertainty or complexity or both.

A typical interpretation is that D is a human being while P is an inanimate system—a "machine" or process (in a wider sense)—for whose proper operation the human being is responsible. The decision maker controls the process using an image P^m to achieve its goal G.

This is also a basic model of an organization. If we compare Fig. 2.2 to Fig. 1.4 of the fundamental organization model, as a whole the combination of the management layer and the operational layer corresponds to D while the process corresponds to P and G corresponds to the organizational goal.

In the human being as a system D is the "brain" of a human being while P is its body. The brain is supposed to have a model P^m of P and a criterion G to evaluate the movement of the body. The specific behavior of the body, which is an external expression of P, is selected for the needs of G. This represents a conventional cybernetic model of a human being.

It must be noticed that in the cybernetic model it does not matter whether or not P^m and G really exist. A goal seeking model is developed and used because it can explain the complex behavior of a human being better than the other models. In other words, P^m and G are used to explain the "logic" of the human behavior rather than the physiological structure. It explains how the system functions rather than what it consists of.

As the above cybernetic interpretation indicates, a goal seeking model presents a model of an internal mechanism of an input output system which from the outside is considered as a black box representation.

Let us formalize Fig. 2.2.

The process of P is an input output system whose input, in general, consists of two components, the external input and the input parameter, m, which is the result of the goal-seeking functioning of D. called decision variable or manipulating variable specified by D. Although the external input of an organization consists of a resource input and an external information input u_e, since the resource input is assumed controllable, it is treated as a part of m. P is then described as having only one internal input m. The resource input is not described in the figure. Notice, however, that the resource input will be treated as an external input in Chapter 11.

Although the information input u_e is supposed to be an input to D, Fig. 2.2 describes that P is affected by u_e as well.

The output of P is y.

The decision maker D, using a feedback information about y and a feedforward information (or an observation) about u_e, selects m as the outcome of solving a decision problem, Δ. The decision problem is based on a model P^m of P and in reference to an assessment (evaluation) G. G is a means to assess desirability of the behavior of P controlled by m.

Let

M = set of alternative decisions,

U_e = set of values of the external input,

and

Y=set of overall outcomes.

Although behavior of an organization member was described in Section 2.1 in terms of an output, y, to a stimulus, u_e, input, a general overall description of the goal seeking system as an input/output system is a pair (y, u_e) which is an element of the Cartesian product of two sets Y and U_e, external input and output pairs, i.e.,

Behavior of goal seeking system = $U_e \times Y$.

The variables u_e, m and y are in general time functions [2] when a goal seeking system is a dynamic system.

Then the process and the assessment function are given by the following two mappings, respectively.

$$P{:}M \times U_e \rightarrow Y$$

and

$$G{:}M \times U_e \times Y \rightarrow Re$$

where Re is the set of reals. Without any constraints we assume in this book that a decision is preferable when it produces a higher value of G. In the case of optimization of G that means maximization.

We shall start conceptualization of a decision problem as maximization for the sake of simplicity. This will be followed by a more general and realistic case, subsequently. Then a basic concept of the goal seeking system is given in terms of the following quintuple:

Definition 2.1 Decision Problem

Given the sets, M, U_e, Y

and the functions

$$P : M \times U_e \rightarrow Y$$

$$G : M \times U_e \times Y \rightarrow R_e$$

The decision problem, Δ, for the optimization case is defined by the statement:

Find m in M such that $G(m,u_e, P(m,u_e))$ is maximized.

The quintuple

$$\Delta = \langle M, U_e, Y, P, G \rangle$$

will be called the structure of a decision problem or just a *decision problem.*

For the notational convenience the following composite representation of G and P, will be used:

$$g: M \times U_e \rightarrow Re$$

defined by

$$g(m,u_e) = G(m,u_e, P(m,u_e)).$$

The corresponding decision problem structure is given by:

$$\Delta = \langle M, U_e, g \rangle.$$

The latter formulation will be referred to as a reduced form.

For the sake of illustration let us consider a typical example of a decision problem given by a dynamic optimization problem [11]. It will be used for the operational level management in Chapter 5.

Let the state equation of a dynamic system be:

$$dz/dt = f(z,m,u_e)$$

where z is a state vector.

Let the evaluation criterion G be given by a functional:

$$r_e = \int_0^T g(z,m,u_e)dt\ .$$

Let the output function be:

$$y = z.$$

The dynamic optimization problem is to find m which maximizes the functional, r_e, for a given u_e. The input output relation of the process P in its evolution over time is given by integration of the state equation.

Although the behavior of a member can be described in the last analysis as an input/output system, to properly understand the behavior and how it can be influenced requires a goal-oriented representation, i.e., in terms of perception, motivation, and goal.

The perception is represented by the observation scheme of u_e. The cognition process should correspond to the process model as a basis for a rational goal seeking activity.

Motivation is represented in terms of the decision problem Δ which the decision maker is using to derive m. A member's behavior is not the result of selecting at random a set of possible alternatives. The member is motivated to select m in accordance with G; that is, the behavior is selected by its motivation to pursue the goal.

The behavior is therefore goal-directed.

It is usual that the needs that direct the motivations of organization member are mutually conflicting. Selfishness and generosity yield a basic cause of conflict. The decision must be made based on some compromise and the compromise is controlled by a decision philosophy, a decision principle, which will be introduced subsequently in the next section. A decision principle can be viewed as exemplifying the personal value system of a member. Every behavior of the goal seeking model should be explained ultimately as conditioned by the value system.

It should be noted that the goal G, which exemplifies the motivation of an organization member, can be considered also as input for the member if it is specified from the outside. This is the case when we discuss the institutional level management in Chapter 10 or garbage can model in Chapter 11.

2.3 Decision Principle and Formal Representation of Goal Seeking System

The above definition of a decision problem has a conceptual difficulty. That is, unless u_e is known, the maximization is not well-defined, i.e., logically impossible to perform because g depends on a specific value of u_e which exists at the time. That implies that g cannot be maximized without knowledge of u_e which by definition is uncertain.

A decision problem is intrinsically indeterminate since it is defined for a set of uncertainty, u_e, while only one uncertainty $u_e \in U_e$ can come to pass at a time. The decision principle makes a decision problem well-defined.

Actually, a decision maker does not make a decision using a decision problem directly. He uses a decision model which is an image of the actual decision problem, i.e., the perception of the problem he is confronted with. Then, a decision maker D is characterized by a pair consisting of a decision model and a decision principle.

A decision model can be anything that can be used to determine a value for m using the decision principle. In order to avoid too much generality, however, the following form of a decision model Δ^m will be used in this book.

$$\Delta^m = \langle M, U_e, Y, P^m, G \rangle$$

where it is assumed that M, U_e, Y and G are the same in Δ and Δ^m.

Although

$$P^m : M \times U_e \to Y$$

is assumed to have the same structure in abstract as P does, it may be very different indeed from P in actual specific realizations. In many cases P^m is a simplified image of P. In particular U_e may be completely omitted as in the case of optimization.

The composite goal g^m is defined then as

$$g^m(m, u_e) = G(m, P^m(m), u_e).$$

$\langle M, U_e, g^m \rangle$ is a reduced form of $\langle M, U_e, Y, P^m, G \rangle$.

When U_e is not a singleton set, a decision maker D has a conceptual difficulty in decision making. That is, if an output cannot be uniquely determined due to existence of U_e, there is no logical base to select a value m

$\in$ M as an optimum decision. The optimization concept, which is the most widely used decision concept, is not applicable to the problem if there are uncertainties.

With uncertainty a decision principle has to be used which reflects the decision making philosophy of the decision maker. In Fig. 2.2 a decision principle is represented by DP.

In the previous section the decision principle was mentioned as a representation of the personal value system. Let us formalize a decision principle.

Let a binary relation $\leqq \subset M \times M$ be called a *preference relation* when it satisfies

(1) $(\forall\ m)(m \leqq m)$

(2) $(\forall\ m,m',m'')(m \leqq m' \text{ and } m' \leqq m'' \rightarrow m \leqq m'')$.

A preference relation is a weak order relation. When $m \leqq m'$ and $m' \leqq m$ hold, we will write

$$m \sim m'$$

and say that m and m' are indifferent to each other.

If $m \sim m'$ implies $m = m'$, the order relation is called a partial order. If we consider the quotient set $M/\sim$ and define the relation on the set as

$$[m] \leqq [m'] \leftrightarrow m \leqq m',$$

the relation is easily proven to be a partial order on the quotient set. In this sense we will use conveniently the term of partial order for the weak order.

When $m \leqq m'$ holds but $m' \leqq m$ does not, we write

$$m < m'$$

and say that m' is strongly preferred to m.

Definition 2.2 Decision Principle

Let $<M, \leqq>$ be a preference relation on M. Then a mapping

$$DP: <M, U_e, g^m> \mapsto <M, \leqq>$$

is called *decision principle.*

$DP(<M,U_e,g^m>)$ is a preference order on M. DP maps a decision model to a preference relation $\leqq$ on M. Since DP is an expression of a philosophy to represent the propensity of the decision maker, DP is naturally required to be applicable to a class of decision models rather than a specific decision model, i.e.,

$$DP:\{<M,U_e,g^m>\} \rightarrow \{<M,\leqq>\}.$$

The domain of a decision principle is a class of decision models. However, sometimes it is extended to include a class of decision problems because a formal structure of a decision problem is the same as that of a decision model and conceptually a decision problem can be a decision model.

A typical example of DP is the max-min decision principle. Let $\leqq_{mm} \subset M \times M$ be:

$$(m,m') \in \leqq_{mm} \leftrightarrow \min_{u_e} g^m(m,u_e) \leqq \min_{u_e} g^m(m',u_e).$$

Then $\leqq_{mm}$ is the order generated by the max-min decision principle. A maximal element of the relation $\leqq_{mm}$ is a max-min solution of $g^m(m,u_e)$. This principle is well known as a precautionary principle [2].

The most important decision principle for the organization theory, which is proposed by Ref. [12], comes from 'bounded rationality' or the satisficing management behavior. The satisficing concept says that what a manager does in practice is not to pursue optimization of his goal but to find a "good enough" solution, i.e., a solution that produces a result better than a pre-assigned level which is called the aspiration level. This management behavior is named satisficing.

The satisficing behavior is formalized as the satisfaction decision principle [2].

Let

$$T:U_e \rightarrow Re$$

be a tolerance function which corresponds to an aspiration level. Then, $m^* \in M$ is called satisfactory if it satisfies the following relation:

$$(\forall u_e)(g^m(m^*,u_e) \geqq T(u_e)).$$

The satisfactory concept divides M into the two categories, satisfactory and unsatisfactory (acceptable and unacceptable). Let

$$M=M_s \cup M_u$$

where M_s is the set of satisfactory elements while M_u is that of unsatisfactory. Let $\leqq_s \subset M \times M$ be a linear order on M such that

$$m <_s m' \leftrightarrow m' \in M_s \text{ and } m \in M_u.$$

Then,

Definition 2.3 Satisfactory Decision Principle
The *satisfactory decision principle* DPs is given as:

$$DPs(<M,U_e,g^m>)=<M,\leqq_s>.$$

Fig. 2.3 shows a concept of the satisfactory decision principle.

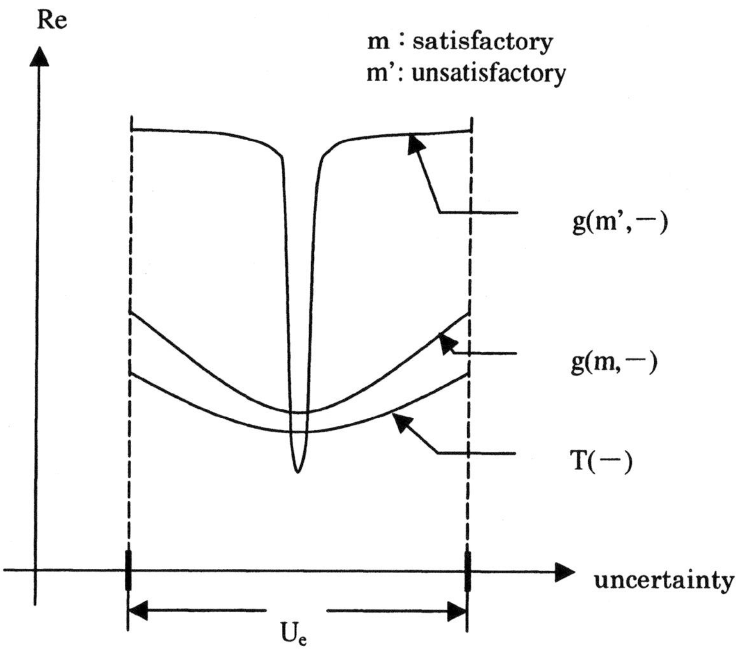

Figure 2.3 Satisfaction decision principle

The decision m is satisfactory because it yields a better result than the tolerance limit $T(u_e)$ for any $u_e \in U_e$. But the decision m' is unsatisfactory because it violates the tolerance limit for some $u_e \in U_e$ although it yields better results than m for almost every $u_e \in U_e$. In this sense the satisfactory decision principle is another precautionary principle [2].

An organization decision problem is almost always multi-objective. Since the multi-objectiveness is an expression of the uncertainty associated motivation, it can be readily reformulated as a decision problem with uncertainty. This straightforward fact is not well recognized.

Let the goal G' of a multi-objective decision problem be

$$G':M \times U_e \times Y \rightarrow Re^n$$

where n is an integer.

Let

$$G'(m,u_e,y)=(G'_1(m,u_e,y),..,G'_n(m,u_e,y))$$

where $G'_i:M \times U_e \times Y \rightarrow Re$.

Let

$$N=\{1,..,n\}.$$

Let

$$G'':M \times U_e \times N \times Y \rightarrow Re$$

be

$$G''(m,u_e,i,y)=G'_i(m,u_e,y).$$

Then, we have a decision problem with an uncertainty input set $U_e \times N$.

The uncertainty associated with a motivation or a multi-objective decision problem also requires a decision principle, i.e., a philosophy for decision making. A typical example of the decision principle for it is the linear weighted sum decision principle. Suppose U_e can be ignored. Let

$$w=(w_1,..,w_n)$$

where $w_i \geqq 0$. w is called the weighting vector which reflects the value system of a decision maker. Then, the relation $\leqq_l \subset M \times M$ which is defined by

$$(m,m') \in \leqq_l \leftrightarrow \sum w_i * g_i^{m'}(m) \leqq \sum w_i * g_i^{m'}(m')$$

specifies a decision principle where $g^{m'}(m)=G_i'(m,P^m(m))$. This principle is called linear weighted sum.

The multi-objective decision making has a special importance for the organization theory. When organization members are represented by goal seeking models, the organization is intrinsically multi-objective. In the subsequent chapters the multi-objective situation will be investigated after it is transformed into a decision problem with uncertainty in the way discussed above.

A goal seeking system is now defined as a combination of a decision model and a decision principle.

Definition 2.4 Goal Seeking System (GSS)

A formal representation of a *goal seeking system* is given by

$$GSS=\langle M,U_e,Y,P,P^m,G,DP\rangle.$$

DP is a principal target of theoretical investigations of a goal seeking system [2] where specific properties are imposed on DP to explore a goal seeking system. For instance, in Chapter 8 we will introduce a function BDP (binary decision principle). BDP is a type of satisfactory decision principle. Two conditions, monotonicity and weak veto power are imposed to examine a BDP interpretable organization.

2.4 Input Output Model Associated With Goal Seeking System: Standard Operating Procedure

As mentioned in Sec. 2.2, a goal seeking system is an internal structure to explain the external behavior of an organization or its members. Let us investigate the relationship between the internal structure of a goal seeking model and the external behavior using the above formulation.

As pointed out, there are two inputs to the goal seeking system: The external input and the goal. Let us consider first the case where the input is the external one and the goal input is fixed.

Let us start with the simplest case where the external input u_e is observable or it can be determined before making a decision.

Then, a value of the decision variable m is determined based on the observation of u_e. Let this alternative be given by a function:

$$\delta:U_e \rightarrow M.$$

δ is usually called *decision function*. This is a representation of a standard operating procedure of the bureaucratic model of organization considered in Chapter 10 [18].

An optimum decision function δ^* can be defined by the condition:
For all u_e

$$\delta^*(u_e)=m^* \rightarrow (\forall\ m)(g^m(m,u_e) \leqq g^m(m^*,u_e)).$$

If we have an optimum decision function, the external behavior of a goal seeking system is given by:

$$P^*:U_e \rightarrow Y$$

such that

$$P^*(u_e)=P(\delta^*(u_e),u_e).$$

The following relation

$$\{(u_e,P^*(u_e))|u_e \in U_e\}$$

is also an external behavior representation of a goal seeking system.

If the composite goal function g^m of the organization does not change, the standard operating procedure δ^* provides an efficient process for decision making because a decision making process can be skipped, i.e., reduced to the application procedure. Furthermore, the operating procedure, by definition, yields an optimum external behavior. But it is well-known that serious problems can exist in a bureaucratic organization; once δ^* is determined, the organization does not change the operating procedure even in the face of changed circumstances, i.e., even if g^m changes.

In reality it is usually the case that u_e is not observable in a precise way. Let us classify such occasions into three cases.

The simplest case is that U_e is a singleton set, that is,

$$U_e=\{u_{e0}\}$$

for some u_{e0}. Then the external behavior is constant

$$P(\delta^*(u_{e0}),u_{e0})$$

as long as g^m is fixed. This case is essentially equivalent to the observability case because we know a true value of the uncertainty before making a decision.

The second case is that although U_e is not a singleton set, a probability model of U_e is available. This is the most important case of this book. The probability used in this book is a subjective probability [13]. Furthermore, let us assume that some observation mechanism exists for U_e. The mechanism corresponds to the perception of Section 2.1.

Let us assume that the relation between u_e and its observed value x is given by a conditional probability.

Let

X=the set of observed values of U_e.

Let the conditional probability be given as:

$Pr(u_e|x)$=conditional probability of u_e on x.

Furthermore, for the sake of simplicity let U_e and X be finite, i.e.,

$$U_e=\{u_{e1},..,u_{el}\} ,$$

$$X=\{x_1,..,x_m\} .$$

Since u_e is observed as x, the decision function concept can be used for x. Let a decision function on X be:

$$\delta_v: X \rightarrow M.$$

Then, the optimum decision function δ_v^* of δ_v is specified by the following condition:

$$\delta_v^* \text{ is optimum} \rightarrow E_{u_e} E_{x|u_e}\, g(\delta_v(x),u_e) \text{ is maximized by } \delta_v^* .$$

E_{ue} and $E_{x|ue}$ are expectation operators with respect to the probability distribution of u_e and the conditional probability distribution of $x|u_e$, respectively. Notice that not g^m but g is used for definition of δ_v^*.

The following fact holds for the optimum δ_v^*.

Proposition 2.1

Suppose $Pr(u_e|x)$ is given. Suppose $P=P^m$. Then, an optimum $\delta_v^*:X \rightarrow M$ is given by the following condition:

$$\delta_v^*(x) = m^* \rightarrow \sum_i g^m(m, u_{ei})Pr(u_{ei} / x) \text{ is maximized by } m^*.$$

The proof is given in Appendix.

The proposition says that if $P=P^m$, an optimum decision is given by maximizing $\Sigma_i\ g^m(m,u_{ei})Pr(u_{ei}|x)$ with respect to m when x is observed. Since $Pr(u_{ei}|x)$ is the probability of u_{ei} for a given x, the statement of the proposition is compatible with our intuitive expectation.

This probabilistic model provides a basis of the adaptation management of the institutional level of this book. Fig. 2.4 shows a conceptual representation of the mechanism of the probabilistic case.

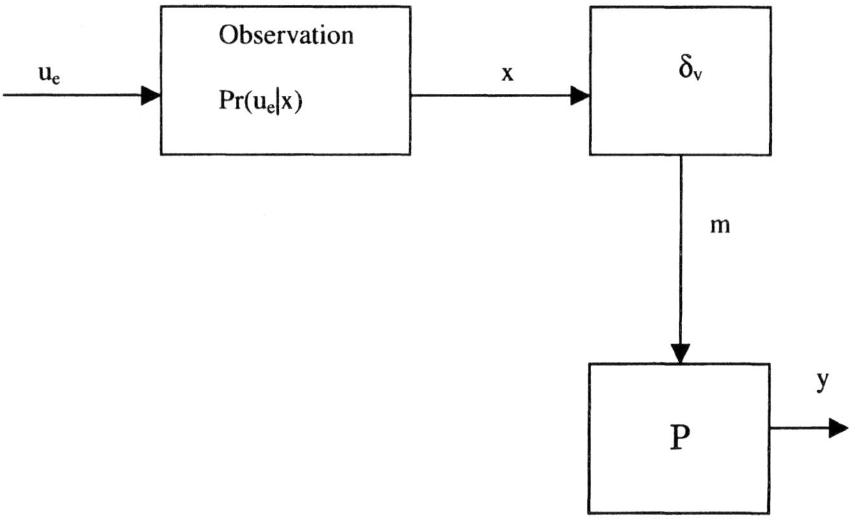

Figure 2.4 Probabilistic case

An ideal adaptive management function of the institutional and the organizational levels is to provide δ_v^* at any time following the change of the environmental structure. Notice that δ_v^* depends on the conditional probability. The process P of Fig. 2.4 corresponds to the overall normative behavior of the operational level in the investigation of the adaptive function of this book.

When a probability model is used for U_e, the external behavior of a goal seeking model is not given by a deterministic function but by a conditional probabilistic relation as follows:

$$\Pr(y|u_e)=\Pr(\{x|y=P(\delta_v^*(x),u_e)\}).$$

Third case is that the only available information is is that the uncertainty will be within U_e and so a decision principle is used.

Let DP be a decision principle adopted. Then, $DP(<M,U_e,g^m>)$ specifies an order relation on M.

Let

$$\sup DP(<M, U_e,g^m>)=m^*$$

where the sup operation is performed with respect to the order on M and m^* is a maximal element.

When a DP is used, the external behavior $P^*:U_e \rightarrow Y$ of a goal seeking model is given by

$$P^*(u_e)=P(m^*,u_e).$$

Notice that m^* is fixed.

So far the goal has been assumed as a fixed parameter. Now let us consider the case where the goal can be an input variable. That is, the means of a higher level management are transformed into ends of a lower level management. The case of the normative behavior of the operational level which will be discussed in Chapter 9 is an example of such a situation. The institutional level, which is the higher level management, uses the global goal generated from the organizational goal adapting to change of the environment as means to control the primary activity of the organization. The managerial and operational levels, which correspond to the lower level management, control the process using the global goal as the ends of their actions.

Since the present situation is valid only for modeling of the managerial and operational levels in this book and since they are assumed closed, the external input u_e for the model is fixed as u_{e0}. For $G:M\times\{u_{e0}\}\times Y\rightarrow Re$ let

$$\underline{\underline{G}}(m,u_{e0},G,y)=G(m,u_{e0},y).$$

Since G is observable because it is a directive, a decision function

$$\delta_G:\{G\} \rightarrow M$$

is applicable to the present case where {G} is the set of directives. Using the similar argument as used in the observable case, the external behavior

$$P_G^* : \{G\} \rightarrow Y$$

is then given by

$$P_G^*(G) = P(\delta_G^*(G), u_{e0})$$

where δ_G^* is an optimum decision function whose formal definition is the same as that of δ_v^* mentioned above.

APPENDIX OF CHAPTER 2

Proof of Proposition 2.1

Suppose δ_v^* satisfies the condition of the proposition. Letδ_v':X → M be an arbitrary. Then,

$$E_{u_e}E_{x|u_e}g^m(\delta_v'(x),u_e) = \sum_i\sum_j g^m(\delta_v'(x_j),u_{ei})\Pr(u_{ei})\Pr(x_j\,|\,u_{ei})$$
$$= \sum_j(\sum_i g^m(\delta_v'(x_j),u_{ei})\Pr(u_{ei}\,|\,x_j))\Pr(x_j)$$

(where the following Bayes relation is used:

$$\Pr(u_e,x) = \Pr(u_e\,|\,x)\Pr(x) = \Pr(x\,|\,u_e)\Pr(u_e))$$
$$\leqq \sum_j(\sum_i g^m(\delta_v^*(x_j),u_{ei})\Pr(u_{ei}\,|\,x_j))\Pr(x_j)$$

(because $\Sigma_i\, g^m(\delta_v'(x_j),u_{ei})\Pr(u_{ei}|x_j) \leqq \Sigma_i\, g^m(\delta_v^*(x_j),u_{ei})\Pr(u_{ei}|x_j)$ for any x_j due to the condition of the proposition)

$$= E_{u_e}E_{x|u_e}\, g^m(\delta_v^*(x),u_e)\,.$$

Q.E.D.

CHAPTER 3 GOAL COMPATIBILITY

Chapter 2 introduced the notion of a goal seeking system as the basic concept in representation of organizations. The essence of a goal seeking system is that its input-output behavior is explained in terms of the pursuance of a goal. An organization itself is apparently a goal seeking system with a stated organizational goal. The goal-seeking paradigm is used repeatedly to describe the behavior of the organization as such as well as in representation of the components of the organization and its members as well. The ultimate objective of a (rational) management is to generate proper operational goals for its constituents consistent with the organizational goal. This has to be accomplished while adapting to changes in the environment (external inputs) and coordinating operational levels (primary activity of an organization) to secure the proper functioning for achieving respective operational goals.

It is natural to expect that there is incompatibility—at least to a degree—between the members' goals and the organizational goal. If the organizational goal is completely in conflict with the members' goals, it is certainly impossible to achieve a successful coordination, i.e., secure the attainment of the organizational goal. This is clearly a conflict situation. This chapter presents a formalization of concepts in such conflicts. The conflict resolution among a range of goals and different levels in an organization is one of the principal problems of the organization theory.

There is another serious conflict in an organization; the conflict between (dynamics of) the environment and adaptation ability of the organization (on the institutional level). Unless some consistency exists between the environment and the adaptive goal, the organization cannot successfully adapt, i.e., function properly. This is the main topic of Chapter 10.

This chapter will define the goal compatibility in terms of decision principles as used throughout the organization. The class of members' goals will be represented as multi-objective. A decision principle is then defined as a mapping from a decision model with a multi-objective goal to an ordered set of organization alternatives. If the order is consistent with (or homomorphic to) the order specified by the organizational goal, the class of the members' goals and the organizational goal are said to be compatible with respect to the decision principle. The compatibility preserving decision principle is called organizational decision principle. If the organization decision principle satisfies a specific property, the organization will be considered to have a desirable organization culture.

This chapter first formalizes the compatibility concept. An intra-level function and the concept of organizational culture are introduced. Next a famous result of the social choice theory is presented as an example of how the compatibility can be satisfied where the desirability of the decision principle is defined by political acceptability. Finally, two basic properties are introduced to characterize a meaningful decision principle for an organization. They will characterize a desirable compatibility.

The subsequent chapters will show that existence of a desirable goal compatibility is an essential condition for realization of a valid organizational structure discussed in Sec. 1.4.

3.1 Goal Compatibility in Decision Principle

a. General concept of goal compatibility

Goal compatibility of an organization is defined as the compatibility between the global goal of the management level and the goals of the operational units. The global goal is derived by adaptation from the organizational goal based on the observed external information inputs, as will be discussed in detail in Chapter 10. It is intuitively clear that a desirable coordination of activities of the operational level can be realized only when the global goal is consistent (compatible) with the goals of the operational level. The compatibility between them is one of the most important issues for successful management.

In order to develop a mathematical consideration of the compatibility, the first task is, naturally, to formalize the compatibility concept.

An organization is a multi-level goal-seeking system with a multiplicity of goals. There are two categories of goals: First, on the one hand, each of the organization members has its own goal specified by its responsibility in an organization, and on the other hand, its own personal value system. The responsibility of the members on the first level is to meet the requirements of the clients/consumers, i.e., to make sure that the basic organizational process is functioning satisfactorily. The responsibility of the members on the higher levels is to influence/motivate the members under its jurisdiction to perform the tasks they are assigned. A basic task of the members on the higher management level is therefore to coordinate the functioning of the members on the lower level so that they perform the required function assigned by the organization while being motivated to do so. The members on the higher level therefore have two goals: First, assigned to them per their position in the organization; and second, the goals of the coordination, i.e., influence the members on the lower level to perform the function required by the organization.

The situation is illustrated in Fig. 3.1 for a simple two-level organization.

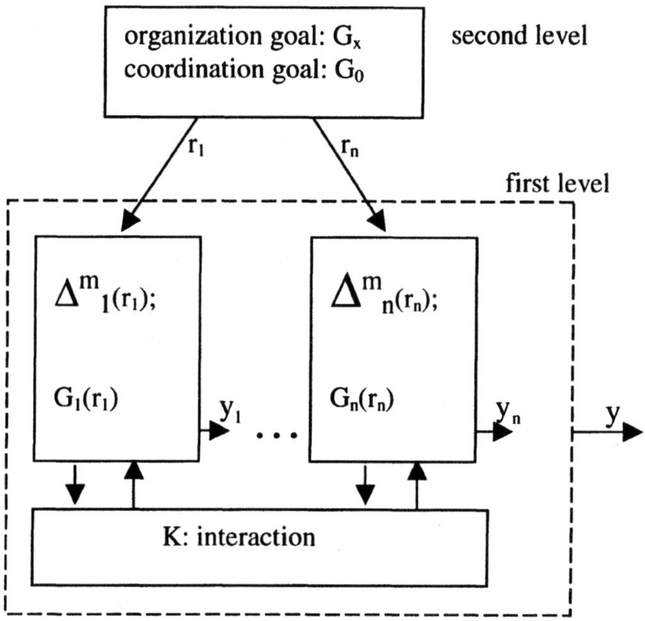

Figure 3.1 Two level organization

The decision model $\{\Delta^m{}_i(r_i)\}$ of the first level members is influenced by the coordination $\{r_i\}$ provided by the second level. The second level has two goals, G_x and G_0: The first refers to satisfying the requirements as assigned by its position in the organization and the second is to influence the members under its jurisdiction to perform the required functions as assigned by the organization. This internlevel relationship replicates throughout a multilevel organization. On the higher level then are an organization goal and a coordination goal.

We can now formalize the relationships between goals within an organization.

Definition 3.1 Organizational goal

Let U_e be the uncertainty set which represents the perception of the organization about the environment and Y the output of the organization, i.e., whatever products or services it is provided. The *organizational goal* is specified by the following mapping:

$$G: U_e \times Y \to Re$$

where Re is Reals.

This definition shows that the organizational goal is "market" driven—as represented by a dependence on the output without including internal manipulating (decision) variables. This view is in agreement with the prevailing view in the organization theory. In this book the organizational goal is always assumed as given by Def. 3.1. We assume further that the uncertainty U_e is characterized by subjective probabilities, $Pr(u_e|x)$ and $Pr_x(x)$ where x is an observed value of $u_e \in U_e$. The goal, i.e., assessment function then depends on the subjective probabilistic at the time.

As Sec. 10.1 will show, the actual organizational goal will have to change in order for the organization to adapt to a changing environment. This results in the fact that the global goal G_x is specified by the adaptation of G to changes in the environment. Formally, let G_x be defined by:

$$G_x{:}Y \to Re.$$

and the goal of the i-th operational unit by:

$$G_i{:}Y_i \to Re\ .$$

That is, an operational unit is also assumed to be concerned only with desirability of its outcome.

The *operational level goals set* is defined by

$$\underline{G}=(G_1,..,G_n)\ .$$

The compatibility problem will be formalized in this chapter in terms of the relationship between the global goal and the operational level goals set $\underline{G}$. The organization overall process output is an n-tuple of outputs of all operational units where Y is the Cartesian product $y \in Y$:

$$Y=Y_1 \times \cdots \times Y_n.$$

Let Π_i be the projection operator, i.e.,

$$\Pi_i(y_1,..,y_i,..,y_n)=y_i$$

and

$$g^h(y,i)= G_i(\Pi_i(y))\equiv g_i(y)$$

g^h will be called *integrated operational goal.*

We then have the following:

Definition 3.2 Multi-objective operational decision problem
Multi-objective operational decision problem associated with $\underline{G}$ is defined by the triplet

$$\Delta(\underline{G})=<Y,N,g^h>$$

The relationship between the plethora of goals in an organization is illustrated in Figure 3.2.

The starting point is the organizational goal G. This is a strategic concern. In adjusting to the prevailing conditions, the organizational goal is modified on the basis of anticipation derived from the observation to produce a "tactical" global goal for the organization, G_x. The task of the management then is to guide the organization to pursue the global goal.

On the lowest level operational units have their own goals, G_i, reflecting their assigned tasks in the organization (Refer to Chapter 7) as well as their personal value systems (Refer to Chapter 8). These form the integrated operational goals set g^h.

There is a gap beween G_x and g^h. To address this problem the coordination goal, G_o, is added to the higher level management. Coordination activity modifies/motivates the individual operational goals, G_i, to pursue, collectively, the global goal of the organization.

The compatibility problem is reduced to resolution of the conflict between the global goal, G_x, and the integrated operational goal, g^h. Management of that conflict is referred to as **coordination** (harmonization).

The figure also shows the relationship among the various goals defined in the organizational model considered here.

As mentioned, the compatibility concept will be formalized using a decision principle. In this book the order relation generated by a decision principle DP will be denoted by $\leqq_{DP(\Delta)}$ (or $<_{DP(\Delta)}$) where Δ is a decision problem.

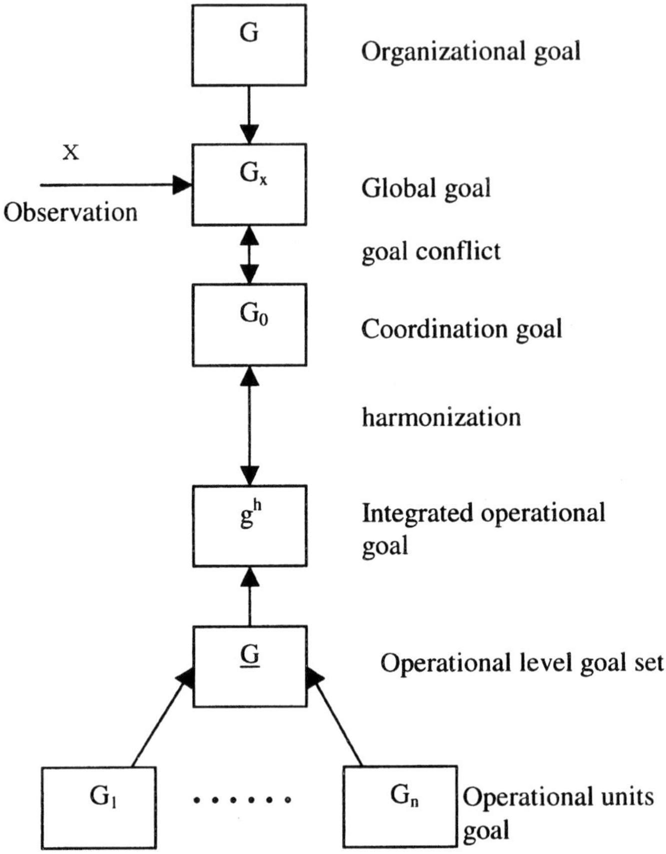

Figure 3.2 Compatibility between global goal and integrated operational level goal

Definition 3.3 Compatibility between G_x and $\underline{G}$

The global goal G_x is said to be *compatible* with the operational level goal set $\underline{G}$ if and only if there exists a decision principle DP

$$DP:\Delta(\underline{G})(=<Y,N,g^h>) \mapsto <Y, \leqq_{DP(\Delta(\underline{G}))}>$$

such that the following relation holds:

$$G_x(y) \leqq G_x(y') \rightarrow y \leqq_{DP(\Delta(\underline{G}))} y'.$$

A DP that realizes the compatibility will be called organizational decision principle and an organization with an organizational decision principle DP will be called DP interpretable organization.

It should be noticed that the organizational decision principle presents an interpretation of the value system of the operational units and its compatibility holds if the preference of the management level as interpreted by operational level is acceptable. Obviously, there are good interpretations and bad interpretations. A good interpretation leads to a successful coordination. It will be characterized later. Notice that

$$y \leqq_{DP(\Delta(\underline{G}))} y' \rightarrow G_x(y) \leqq G_x(y')$$

is not required in the above definition. In general, the order generated by G_x is finer than the outcome of DP when the compatibility holds.

The compatibility concept implies the following definition.

Definition 3.4 Goal integration

$\underline{G}$ is said to be *integrable* to G_x if and only if there exists DP such that G_x is compatible with $\underline{G}$ with respect to DP.

The following proposition is trivial but conceptually important.

Proposition 3.1

For any global goal and operational level goals there exists an organizational decision principle so that the global goal can be made compatible with the operational level goals.

The proof is given in Appendix

Prop. 3.1 implies that an existence problem of DP is not important but its properties are. Actually, it will be shown that a successful coordination can be achieved only if DP satisfies some required conditions which indicate that DP is a good organizational decision principle. A successful integration depends on the properties of DP.

Next, we will introduce two aspects of compatibility—intra-level function and organizational culture—as structural organizational decision principles.

b. Compatibility in Intra-level function

We first introduce the concept of intra-level function.

Definition 3.5 Intra-level function

Suppose $G_x(y)$ and $(G_1(y_1),\ldots,G_n(y_n))$ are related by a function

$$F{:}Re^n \rightarrow Re$$

such that

$$G_x(y)=F(G_1(\Pi_1(y)),\ldots,G_n(\Pi_n(y))).$$

Then, F is called *intra-level function.*

The concept of intra-level function might seem artificial. But there are many cases where G_x is designed or $\underline{G}$ is modified for a good coordination. Since compatibility is necessary for successful coordination, G_x and $\underline{G}$ are designed so that an intra-level function can exist. Chapter 7 will discuss this case.

The following proposition shows that an intra-level function is a structural requirement for an organizational decision principle.

Proposition 3.2
If G_x and $\underline{G}$ are related by an intra-level function F, then G_x is compatible with $\underline{G}$.

The proof is given in Appendix.

c. Compatibility in organizational culture
Let

$$\underline{\Delta}= \text{the class of } \Delta(\underline{G}).$$

Definition 3.6 Target outcome of global goal
Let $G_x{:}Y \rightarrow Re$ be the global goal. Then, let

$$Opt{:}G_x \mapsto Y' \subset Y$$

be

$$Opt(G_x)=\{y|(\forall y')(G_x(y') \leqq G_x(y))\}.$$

The function Opt yields the set of optimum outcomes for each global goal. $Opt(G_x)$ is called *target outcome of the global goal G_x*.

Definition 3.7 Organizational culture
A satisfactory decision principle

$$BDP:\underline{\Delta} \rightarrow \wp(Y)$$

is called *organizational culture* if

$$(\forall i)(g_i(y) \leqq g_i(y')) \rightarrow y \leqq_{BDP(\Delta(\underline{G}))} y'$$

holds.

First let us notice that BDP is a decision principle in the following sense: Let $BDP(\Delta(\underline{G})) = Y'$ and let an order relation $\leqq_{BDP(\Delta(\underline{G}))}$ on Y be such that: for any y and $y' \in Y$

$$y \leqq_{BDP(\Delta(\underline{G}))} y' \leftrightarrow (y \notin Y' \text{ or } y' \in Y')$$

and

$$y <_{BDP(\Delta(\underline{G}))} y' \leftrightarrow y \notin Y' \text{ and } y' \in Y'.$$

Then, BDP can be interpreted as a decision principle given by

$$BDP(\Delta(\underline{G})) = \langle Y, \leqq_{BDP(\Delta(\underline{G}))} \rangle .$$

$\leqq_{BDP(\Delta(\underline{G}))}$ is a binary order relation which has only the maximum (satisfactory) elements and the minimum (unsatisfactory) elements. $y' \in Y'$ and $y \notin Y'$ are maximum and minimum element, respectively.

$$BDP(\Delta(\underline{G})) = Y' \subset Y$$

is interpreted as a set of outcomes which can be accepted by the operational level units as a satisfactory and hence they are supposed to cooperate for producing the output. In this sense BDP reflects the culture of an organization.

Definition 3.8 Compatibility with respect to organizational culture

The global goal G_x is said to be *compatible* with respect to an organization culture $BDP:\underline{\Delta} \rightarrow \wp(Y)$ for $\underline{G}$ if BDP satisfies the following relation:

$$BDP(\Delta(\underline{G})) = Opt(G_x) .$$

In particular, a BDP which satisfies the above relation will be called a desirable organization culture and the organization is also considered to have a desirable organization culture characterized by BDP.

Definition 3.9 BDP interpretable organization

An organization is called *BDP interpretable* if there exists an organizational culture BDP:$\underline{\Delta} \rightarrow \wp(Y)$ such that its global goal and operational level goals are always compatible with respect to the BDP, i.e.,

$$\mathrm{BDP}(\Delta(\underline{G})) = \mathrm{Opt}(G_x).$$

In reality the global goal and the operational level goals can change dynamically but the relationship between them is assumed to remain constant for a BDP interpretable organization. We have the following.

Proposition 3.3

A BDP interpretable organization is a DP interpretable organization. That is, if an organization has a desirable organization culture of BDP, G_x is compatible with $\underline{G}$.

The proof is given in Appendix.

Chapter 9 will discuss a BDP interpretable organization.

3.2 Goal Compatibility Example: Political Desirability of Decision Principle

According to the definition of the compatibility the global goal is said to be compatible with the operational level goals if their relation is explained by a decision principle. Prop. 3.1 says, however, that we can always find a DP which makes the compatibility satisfied if we ignore properties of DP. Whether or not the compatibility is meaningful, then, depends on the properties of the decision principle. Ref. [2] presents three desirable, meaningful properties—Pareto consistency, similarity condition and invariance condition—in order to characterize a rational decision making behavior. The first two properties will be introduced in Sec. 3.3 as necessary for a desirable compatibility for an organization.

This section presents a classical example of the compatibility property and a famous result about that problem [2]. The example is an application of the compatibility concept to a political situation.

For the development of this section three properties—monotonicity, citizen's sovereignty and non-dictatorship—will be introduced as politically acceptable properties for a decision principle. The theoretical result of this section says that the axiom system specified by the assumed properties is inconsistent, that is, there is no decision principle which satisfies all three properties. Consequently, a global goal does not exist which can be made compatible with the given operational level goals if the necessary DP is required to satisfy the three properties indicated.

Although the result is negative, it is important because the example presents a basic theoretical scheme for the compatibility argument.

Let $\underline{D}$ be the class of operational decision models, i.e.,

$$<Y,N,g^h> \in \underline{\Delta}.$$

Let $\underline{L}$ be the class of order relations on outcome sets, i.e.,

$$<Y, \leqq> \in \underline{L}.$$

The following is a technical definition.

Definition 3.10 Privilege

Suppose Y and N are fixed. Then, a subset $K \subset N$ is said to have the *privilege* over the relation $y' < y$ on Y with respect to a decision principle $DP:\underline{\Delta} \to \underline{L}$ if and only if for any $g^h: Y \times N \to Re$ or $\Delta=<Y,N,g^h>$

$$(\forall k \in K)(g^h(y',k) < g^h(y,k)) \to (y' <_{DP(\Delta)} y)$$

holds.

In other words, the order on y, y' is determined solely by opinions of a subgroup of members $K \subset N$. The decision principle, DP, has the property that if the members of K agree that y is preferable to y', DP generates the preference order regardless of opinions of the remaining members N–K.

Lemma 3.1

Suppose a decision principle $DP:\underline{\Delta} \to \underline{L}$ satisfies the following four conditions A.1 to A.4.

***A.1**: Let $\Delta=<Y,N,g^h>$, $\Delta'=<Y,N,g^{h\prime}>$ and $y \in Y$ be arbitrary. Suppose they satisfy the following two conditions:*

(i) For any $i \in N$, y'' and $y''' \in Y$ if $y'' \neq y$ and $y''' \neq y$,

$$g^h(y'',i) \leqq g^h(y''',i) \leftrightarrow g^{h\prime}(y'',i) \leqq g^{h\prime}(y''',i).$$

(ii) For any $i \in N$ and $y'' \in Y$

$$g^h(y'',i) \leqq g^h(y,i) \rightarrow g^{h\prime}(y'',i) \leqq g^{h\prime}(y,i)$$

and

$$g^h(y'',i) < g^h(y,i) \rightarrow g^{h\prime}(y'',i) < g^{h\prime}(y,i).$$

Then, for any $y' \in Y$

$$y' <_{DP(\Delta)} y \rightarrow y' <_{DP(\Delta')} y.$$

A.2: *Let* $\Delta=<Y,N,g^h>$ *and* $\Delta'=<Y,N,g^{h\prime}>$ *be arbitrary. If the following relation*

$$g^h(y,i) \leqq g^h(y',i) \leftrightarrow g^{h\prime}(y,i) \leqq g^{h\prime}(y',i)$$

holds for every y *and* $y' \in Y$ *and* $i \in N$, *then*

$$DP(\Delta)=DP(\Delta').$$

A.3: *Let* $\Delta=<Y,N,g^h>$ *and* $Y' \subset Y$ *be arbitrary. Then,*

$$DP(\Delta|Y')=DP(\Delta) \cap (Y' \times Y')$$

where $\Delta|Y'=<Y',N,g^h|Y' \times N>$.

A.4: *Let* y *and* $y' \in Y$ *be arbitrary. Then, for any* N *there exist* $\Delta=<Y,N,g^h>$ *and* $\Delta'=<Y,N,g^{h\prime}>$ *in* $\underline{\Delta}$ *such that*

$$y <_{DP(\Delta)} y'$$

and

$$y' <_{DP(\Delta')} y.$$

Then, DP satisfies the following relation which will be called Pareto property:

$$(\forall i)(g^h(y',i) < g^h(y,i)) \rightarrow y' <_{DP(\Delta)} y.$$

The proof is given in Appendix.

The conditions A.1 and A.4 are usually called monotonicity and citizen's sovereignty, respectively. A1 considers the situation where g^h changes to $g^{h\prime}$ in the way that for all operational unit the preference relation between $y''\neq y$ and $y'''\neq y$ is not changed and the preference of y is not degraded. Then, the preference of y is preserved against the change. A2 says that the preference relation between y and y' can be controlled by the integrated operational goal g^h or the preferences of the operational units.

Lemma 3.2

Suppose a decision principle $DP:\underline{\Delta} \to \underline{L}$ satisfies A.1, A.2 and A.3 of Lemma 3.1. Let N and $K \subset N$ be given. Suppose there exist some $\Delta = \langle Y, N, g^h \rangle$ and $y, y' \in Y$ such that

$$g^h(y',k) < g^h(y,k)$$

for every $k \in K$ and

$$g^h(y,j) < g^h(y',j)$$

for every $j \in N-K$ and

$$y' <_{DP(\Delta)} y$$

hold. Then, K has the privilege over $y' < y$ on Y with respect to DP.

The proof is given in Appendix.

Proposition 3.4

Suppose $|Y| \geqq 3$. There is no decision principle that satisfies the conditions from A.1 to A.4 of Lemma 3.1 and the following A.5:

A.5: *For any Y and N, and $i \in N$ there exists g^h or $\Delta = \langle Y, N, g^h \rangle$ in $\underline{\Delta}$ and $y, y' \in Y$ such that*

$$g^h(y',i) < g^h(y,i)$$

and

$$y <_{DP(\Delta)} y'.$$

The proof is given in Appendix.

The condition A.5 is called *non-dictatorship* condition. It says that for any person there is at least one pair (y,y') whose preference relation cannot be controlled by his own will.

3.3 Required Properties for Decision Principle

The properties of a decision principle used in Sec. 3.2 are considered desirable from a political perspective and hence the result of Sec. 3.2 has been appreciated as a basic theory of the social choice.

Since an organization is a contrived system with a pre-assigned goal, it is different from a society and hence the three properties cannot be necessarily accepted as desirable. For an organization a style of leadership might be considered important which may be incompatible with the non-dictatorship. The monotonicity, on the other hand, will be shown as essential for a successful coordination. The citizen's sovereignty may be or may not be important for successful management.

The axiom system of Sec. 3.2 neglects a quantitative aspect of $\underline{G}$ and the theory as constructed is concerned only with a qualitative aspect. This fact is typically demonstrated by A.1 and A.3. It is definitely true, however, that quantitative aspects cannot be ignored for evaluation of organization activities.

Let us consider the linear weighted sum decision principle DP_l which generates a typical integrated goal $G^h:Y\rightarrow Re$ from $\underline{G}$. DP_l is an important decision principle in management because it is widely used due to its easy interpretation.

Let G^h be given by:

$$G^h(y)=w_1*g^h(y,1)+\cdots+w_n*g^h(y,n)$$

where $w_i \geqq 0$.

$$w=(w_1,..,w_n)$$

is called weight vector.

DP_l is given by: For $\Delta=\langle Y,N,g^h\rangle$

$$y \leqq_{DPl(\Delta)} y' \leftrightarrow G^h(y) \leqq G^h(y').$$

It is obvious that if $G_x= G^h$ then DP_l is an organizational decision principle.

DP_l does not satisfy A.1 and A.3 in general. For instance, suppose

$$Y= \{y,y'\}$$

and

$N=\{1,2\}$.

Suppose $g_i:Y \rightarrow Re$ and $g_i':Y \rightarrow Re$ are given by the following tables:

g^h	g_1	g_2
y'	0	50
y	100	0

$g^{h'}$	g_1'	g_2'
y'	99	50
y	100	0

Then, for instance, $g_1(y')=0$. Let

$\Delta=<Y,N,g^h>$

and

$\Delta'=<Y,N,g^{h'}>$

and

$w=(1,1)$.

If $y''\neq y$ and $y'''\neq y$, we have $y''=y'''=y'$. Then

$$g^h(y'',i) \leqq g^h(y''',i) \leftrightarrow g^{h'}(y'',i) \leqq g^{h'}(y''',i).$$

The relation

$$g^h(y',i) < g^h(y,i) \rightarrow g^{h'}(y',i) < g^{h'}(y,i)$$

also holds. Since $G^h(y')=50$ and $G^h(y)=100$, $y' <_{DPI(\Delta)} y$. But since $G^{h'}(y')=149$ and $G^{h'}(y)=100$, $y <_{DPI(\Delta')} y'$ which contradicts A.1.

Similarly,

$$g^h(y',i) \leqq g^h(y,i) \leftrightarrow g^{h'}(y',i) \leqq g^{h'}(y,i)$$

holds for all $i \in N$. But $y' <_{DPI(\Delta)} y$ and $y <_{DPI(\Delta')} y'$, which contradicts A.3.

The latter contradiction clearly shows that A.3 neglects the quantitative aspect of $\underline{G}$. Characterization of the linear weighted sum decision principle requires an axiom system different from that of Sec. 3.2 [2].

There may be cases when for an organization only the qualitative properties of $\underline{G}$ are important. In usual cases, however, the quantitative aspect plays a vital role in comparison between $\underline{G}$ and G_x .

We will introduce two properties as desirable ones for an organizational decision principle which can address the quantitative aspect as well as the qualitative aspect of $\underline{G}$.

Definition 3.11Pareto property

A decision principle DP is said to satisfy *Pareto property* if and only if the following relation holds:

$$(\forall\, i \in N)(g^h(y,i) \leqq g^h(y',i)) \rightarrow y \leqq_{DP(\Delta)} y'.$$

The Pareto property requires that the order generated by a decision principle must not be against the desire of the operational level units and hence the global goal which is compatible with the operational level goals with respect to the decision principle must not be against the desire either.

Lemma 3.1 shows that if a decision principle satisfies A.1, A.2, A.3 and A.4, it has the Pareto property. If the Pareto property does not hold, any coordination scheme may not be able to work in the desired way. Def. 3.7 explicitly requires that an organizational culture must satisfies the Pareto property.

The property also holds for DP_l because if $g^h(y,i) \leqq g^h(y',i)$ for all $i \in N$, $G^h(y) \leqq G^h(y')$ since $0 \leqq w_i$, that is, $y \leqq_{DPl(\Delta)} y'$.

Definition 3.12 Similarity condition

Let $\Delta = <Y,N,g^h>$ and $\Delta' = <Y',N',g^{h'}>$ be decision problems. Let

$$h_1 : Y \rightarrow Y'$$

and

$$h_2 : N \rightarrow N'.$$

Suppose the following relation holds:

$$(\forall\, y)(\forall\, i)(g^{h'}(h_1(y),h_2(i)) = g^h(y,i)) \rightarrow (y \leqq_{DP(\Delta)} y' \rightarrow h_1(y) \leqq_{DP(\Delta')} h_1(y')).$$

Then, DP is said to satisfy *similarity condition.*

Since $(h_1,h_2):\Delta \rightarrow \Delta'$ can be considered as a homomorphism from Δ to Δ' (Fig. 3.2 is commutative), the similarity condition requires that a meaningful decision principle should produce a similar order for a similar decision problem.

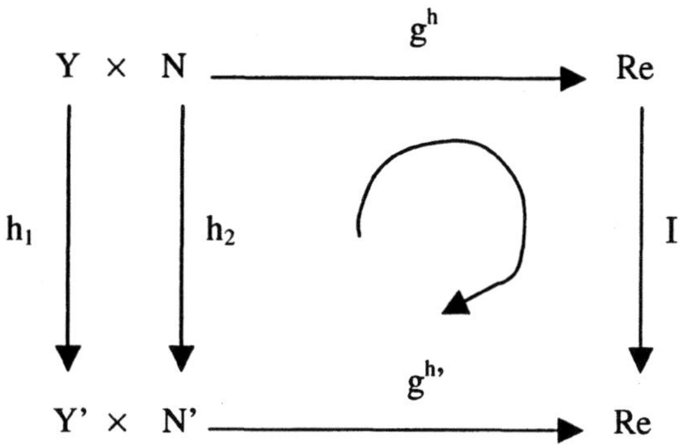

Figure 3.3 Homomorphic relation between decision problems

Lemma 3.3

A.2 implies the similarity condition where $h_1=I$ and $h_2=I$.

The proof is given in Appendix.

If the similarity condition holds, the principle of policy continuity is satisfied for the goal integration. This is important for a stable organizational culture.

APPENDIX OF CHAPTER 3

Proof of Proposition 3.1

Let a decision principle DP be: For any decision problem $\Delta=<Y,N,g^h>$

$$DP(\Delta)=<Y, \leqq_{DP(\Delta)}>$$

where

$$\leqq_{DP(\Delta)} = Y \times Y$$

or there is no preference among y's. Then, let

$$G_x(y) \leqq G_x(y') .$$

If y $\leqq_{DP(\Delta)}$ y' is not true,

$$y >_{DP(\Delta)} y'$$

should hold, which contradicts $(y,y') \in \leqq_{DP(\Delta)}$.

Q.E.D.

Proof of Proposition 3.2

Suppose G_x and $\underline{G}$ are related as

$$G_x(y)=F(g_1(y),\ldots,g_n(y)).$$

Let a decision function DP be induced by F as follows:

$$y \leqq_{DP(\Delta(\underline{G}))} y' \leftrightarrow F(g_1(y),\ldots,g_n(y)) \leqq F(g_1(y'),\ldots,g_n(y')).$$

Then, the following is trivially true:

$$G_x(y) \leqq G_x(y') \rightarrow y \leqq_{DP(\Delta(\underline{G}))} y'.$$

That is, G_x is compatible with $\underline{G}$ with respect to DP.

Q.E.D.

Proof of Proposition 3.3

Suppose an organization is a BDP interpretable, i.e.,

$$BDP(\Delta(\underline{G})) = Opt(G_x) .$$

We will show that

$$G_x(y) \leqq G_x(y') \rightarrow y \leqq_{BDP(\Delta(\underline{G}))} y'.$$

If $y >_{BDP(\Delta(\underline{G}))} y'$, the following is true:

$$y \in BDP(\Delta(\underline{G}))$$

and

$$y' \notin BDP(\Delta(\underline{G}))$$

due to the definition of $BDP(\Delta(\underline{G}))$. Then,

$$y \in Opt(G_x)$$

holds due to the assumption. Since $G_x(y) \leqq G_x(y')$, then

$$y' \in Opt(G_x)$$

that is,

$$y' \in BDP(\Delta(\underline{G})),$$

which is a contradiction.

Q.E.D.

Proof of Lemma 3.1

Suppose y and y' are fixed. Suppose

$$g^h(y',i) < g^h(y,i)$$

for all $i \in N$. We will show $y' <_{DP(\Delta)} y$.

The condition A.4 implies that there exists $\Delta'=<Y,N,g^{h'}>$ in $\underline{\Delta}$ such that $y' <_{DP(\Delta')} y$. Let $\Delta''=<Y,N,g^{h''}>$ be:

$$g^{h''}|(Y-\{y\})\times N = g^{h'}|(Y-\{y\})\times N,$$

$$g^{h''}(y',i) < g^{h''}(y,i)$$

and

$$g^{h'}(y,i) \leqq g^{h''}(y,i)$$

for any $i \in N$.

Let $i \in N$ and $y'',y''' \in Y$ be arbitrary where $y'' \neq y$ and $y''' \neq y$. Then,

$$g^{h'}(y'',i)=g^{h''}(y'',i)$$

and

$$g^{h'}(y''',i)=g^{h''}(y''',i).$$

Hence (i) of A.1 holds.

Suppose

$$g^{h'}(y'',i) \leqq g^{h'}(y,i).$$

Since $g^{h''}(y'',i) = g^{h'}(y'',i)$ and $g^{h'}(y,i) \leqq g^{h''}(y,i)$,

$$g^{h''}(y'',i) = g^{h'}(y'',i) \leqq g^{h'}(y,i) \leqq g^{h''}(y,i)$$

holds. Similarly,

$$g^{h'}(y'',i) < g^{h'}(y,i) \rightarrow g^{h''}(y'',i) < g^{h''}(y,i)$$

can be shown. Hence,(ii) of A.1 is true. Then the condition A.1 implies that $y' <_{DP(\Delta'')} y$.

Let $Y'=\{y,y'\}$. Then the condition A.3 implies

$$DP(\Delta)\cap(Y'\times Y')=DP(\Delta|Y')$$

$$=DP(\Delta''|Y')$$

(due to the assumption $g^{h}(y',i) < g^{h}(y,i)$, the condition $g^{h''}(y',i) < g^{h''}(y,i)$ and A.2)

$$=DP(\Delta'') \cap (Y' \times Y').$$

Consequently, we have $y' <_{DP(\Delta)} y$.

Q.E.D.

Proof of Lemma 3.2

Let y and y' be fixed. Let $\Delta'=<Y,N,g^{h'}>$ be arbitrary where $g^{h'}(y',k) < g^{h'}(y,k)$ for k in K. We will show $y' <_{DP(\Delta')} y$.

Let $\Delta''=<Y,N,g^{h''}>$ be such that

$$g^{h''}|(Y-\{y\}) \times N = g^{h}|(Y-\{y\}) \times N,$$

and

$$g^{h'}(y',i) \leqq g^{h'}(y,i) \leftrightarrow g^{h''}(y',i) \leqq g^{h''}(y,i)$$

for any $i \in N$. Let $Y'=\{y, y'\}$. Let us examine $\Delta|Y'$ and $\Delta''|Y'$.

Let us check the conditions of A.1.

Condition(i)

If $y'' \neq y$, $y''' \neq y$

$$y''=y'''=y'.$$

Then

$$g^{h}(y'',i) \leqq g^{h}(y''',i) \leftrightarrow g^{h''}(y'',i) \leqq g^{h''}(y''',i)$$

is trivially true.

Condition(ii)

Case a: $i \in K$.

In this case the assumption of the lemma implies

$$g^{h}(y',i) < g^{h}(y,i).$$

Since

$$g^{h'}(y',i) < g^{h'}(y,i) \quad \text{due to the definition of } \Delta'$$

and

$$g^{h'}(y',j) < g^{h'}(y,j) \leftrightarrow g^{h''}(y',j) < g^{h''}(y,j)$$

for any $j \in N$ due to the definition of $g^{h''}$,

$$g^{h''}(y',i) < g^{h''}(y,i)$$

holds.

$$g^{h'''}(y',i) < g^{h'''}(y,i) \rightarrow g^{h'''}(y',i) \leqq g^{h'''}(y,i)$$

is also true. Therefore,

$$g^{h}(y'',i) \leqq g^{h}(y,i) \rightarrow g^{h'''}(y'',i) \leqq g^{h'''}(y,i)$$

and

$$g^{h}(y'',i) < g^{h}(y,i) \rightarrow g^{h'''}(y'',i) < g^{h'''}(y,i)$$

hold.

Case b: $i \notin K$.
Since $g^{h}(y',i) > g^{h}(y,i)$,

$$g^{h}(y',i) \leqq g^{h}(y,i) \rightarrow g^{h'''}(y',i) \leqq g^{h'''}(y,i)$$

and

$$g^{h}(y',i) < g^{h}(y,i) \rightarrow g^{h'''}(y',i) < g^{h'''}(y,i)$$

are trivially true (because the left hand side of the implications are false). Therefore, A.1 implies

$$y' <_{DP(\Delta|Y')} y \rightarrow y' <_{DP(\Delta''|Y')} y.$$

Since $y' <_{DP(\Delta)} y$ is true due to the assumption of the lemma,

$$y' <_{DP(\Delta''|Y')} y$$

is true. Consequently, A.3 implies

$$DP(\Delta|Y') = DP(\Delta) \cap (Y' \times Y')$$

$$= DP(\Delta'') \cap (Y' \times Y') = DP(\Delta''|Y').$$

Let us consider $\Delta''|Y'$ and $\Delta'|Y'$. Because of the definition of $g^{h'''}$ A.2 implies

$$DP(\Delta''|Y')=DP(\Delta'|Y').$$

Hence,

$$DP(\Delta)\cap(Y'\times Y')= DP(\Delta|Y')=DP(\Delta''|Y')$$

$$=DP(\Delta'|Y')$$

$$=DP(\Delta')\cap(Y'\times Y').$$

Therefore, since $y' <_{DP(\Delta)} y$

$$y' <_{DP(\Delta')} y$$

is true.

Q.E.D.

Proof of Proposition 3.4

Suppose there were a decision principle $DP:\underline{\Delta} \rightarrow \underline{L}$ which satisfies the conditions A.1, A.2, A.3, A.4 and A.5.

Let y and y' in Y be arbitrary. Notice that N has the privilege over $y' < y$ with respect to DP due to Lemma 3.1. Furthermore, notice that the empty set cannot have the privilege over $y' < y$ because if it could, the complement of the empty set, i.e., N could not. Then, there is a minimal set $N^* \subset N$ which has the privilege over $y' < y$.

Let $i \in N^*$ ($N^* \neq \phi$) be fixed. Let $K=N^* -\{i\}$ and $L=N-N^*$. We will show that $\{i\}$ has the privilege over any $y'' < y'''$, that is, $\{i\}$ violates the condition A.5.

Suppose $y'' \neq y, y'' \neq y'$ and $y''' \neq y$. Let $\Delta=\langle Y,N,g^h\rangle$ be such that

$$g^h(y'',i) <g^h(y',i) <g^h(y,i)\ ,$$

$$g^h(y',k) <g^h(y,k) <g^h(y'',k) \text{ for k in K}\ ,$$

$$g^h(y,l) <g^h(y'',l) <g^h(y',l) \text{ for l in L.}$$

Since $g^h(y',j) < g^h(y,j)$ for $j \in N^*=\{i\} \cup K$ and since N^* has the privilege over $y' < y$, $y' <_{DP(\Delta)} y$ holds.

Let us consider the relation between y' and y". Suppose $y' <_{DP(\Delta)} y''$ is true. Then, since $g^h(y',k) < g^h(y'',k)$ for $k \in K$ and since $g^h(y'',j) < g^h(y',j)$ for $j \in N-K=\{i\} \cap L$, K has the privilege over $y' < y''$ due to Lemma 3.2.

Then, let $\Delta'=<Y,N,g^{h'}>$ be

$$g^{h'}(y'',i) < g^{h'}(y,i) < g^{h'}(y',i),$$

$$g^{h'}(y',k) < g^{h'}(y'',k) < g^{h'}(y,k) \text{ for k in K},$$

and

$$g^{h'}(y'',l) < g^{h'}(y,l) < g^{h'}(y',l) \text{ for l in L}.$$

Since K has the privilege over $y' < y''$, $y' <_{DP(\Delta')} y''$ is true. Lemma 3.1 also implies $y'' <_{DP(\Delta')} y$. Consequently, $y' <_{DP(\Delta')} y$.

On the other hand, since $g^{h'}(y',k) < g^{h'}(y,k)$ for $k \in K$ and since $g^{h'}(y,j) < g^{h'}(y',j)$ for $j \in N-K$, Lemma 3.1 implies that K should have the privilege over $y' < y$ which contradicts with the minimality of N^*. Hence, $y'' \leqq_{DP(\Delta)} y'$ should be true. Consequently, $y'' <_{DP(\Delta)} y$.

On the other hand, since $g^h(y'',i) < g^h(y,i)$ and $g^h(y,j) < g^h(y'',j)$ for $j \in N-\{i\}$, $\{i\}$ has the privilege over $y'' < y$ due to Lemma 3.2. Let $\Delta''=<Y,N,g^{h''}>$ be such that

$$g^{h''}(y'',i) < g^{h''}(y,i) < g^{h''}(y''',i),$$

$$g^{h''}(y,j) < g^{h''}(y''',j) < g^{h''}(y'',j)$$

for j in $N-\{i\}$. Since $\{i\}$ has the privilege over $y'' < y$, $y'' <_{DP(\Delta'')} y$ holds. Then, since $g^{h''}(y,j) < g^{h''}(y''',j)$ for every $j \in N$, $y <_{DP(\Delta'')} y'''$ holds due to Lemma 3.1. Hence, $y'' <_{DP(\Delta'')} y'''$.

Since $g^{h''}(y''',j) < g^{h''}(y'',j)$ for j in $N-\{i\}$, $\{i\}$ has the privilege over $y'' < y'''$.

The other case where $y'' \neq y$, $y'' \neq y'$ and $y''' \neq y$ do not hold can be treated in the same way.

Q.E.D.

Proof of Lemma 3.3

Let $\Delta=<Y,N,g^h>$ and $\Delta'=<Y,N,g^{h'}>$ be decision problems. Let $h_1=I$ and $h_2=I$ be the identities. We have to show that

$$(\forall\, y \in Y)(\forall\, i \in N)(g^{h}(y,i) = g^{h'}(y,i)) \rightarrow (y \leqq_{DP(\Delta)} y' \rightarrow y \leqq_{DP(\Delta')} y').$$

Suppose $g^{h}(y,i)=g^{h'}(y',i)$ for any i and y. Then,

$$g^{h}(y,i) \leqq g^{h}(y',i) \leftrightarrow g^{h'}(y,i) \leqq g^{h'}(y',i).$$

A.2 implies that

$$y \leqq_{DP(\Delta)} y' \leftrightarrow y \leqq_{DP(\Delta')} y'.$$

In particular,

$$y \leqq_{DP(\Delta)} y' \rightarrow y \leqq_{DP(\Delta')} y'$$

holds.

Q.E.D.

CHAPTER 4 ORGANIZATION SKELETON

As the starting point of building our organizational cybernetic theory, this chapter will investigate the skeleton of an organization in the simplest form where only the positions of the organization members in the organizational hierarchy are considered. They are represented simply as points having no structure. Even in this simplest form a hierarchical structure can be found to be fundamental character of an organization. This fact is an illustration to support the thesis that a complex system cannot be managed or controlled without introducing a hierarchical structure. Furthermore, it has been observed that every system has a tendency to increase the height of its hierarchy when it confronts a more complicated situation. This is clear if we notice how a committee produces a subcommittee or a taskforce, and how a newly established organization again establishes sub-organization. Although this tendency may be inevitable, it has been also pointed out that such a tendency decreases efficiency of an organization; for instance, a higher hierarchy leads to slower decision making. On the other hand as the hierarchy of an organization is made flatter, its span of control must become wider or a supervisor may not be able to manage his subordinates effectively due to the increased span of control. A lower hierarchy can be incompatible with realization of an optimum supervising capability. Further studies of the hierarchical structures can be a good research target of the organization theory.

The hierarchical structure is defined in this chapter by supervisory relations among members. This reflects an organization chart. The structure is, mathematically speaking, represented as a simple partially ordered relation but we can develop interesting facts about it by introducing specific conditions which reflect the fact that the target system is not just a simple mathematical object but an organization.

We first formalize the hierarchy of an organization as a tree structure whose top node is the president of the organization and whose leaf nodes represent operators. In this chapter an operational level decision maker will be called operator because he is supposed to carry out an order given by his supervisor faithfully rather than to behave as a goal seeking object on its own. An organization member who is not an operator is called manager. The tree structure representation enables us to formalize classical concepts of an organization, such as span of control, unity of command and others. This chapter does not try to exhaust all classical concepts. Based on the formalized concepts some taxonomy of organization structures can be derived.

Finally, theoretical considerations of an organization chart using the taxonomy can be conducted. The theoretical investigation is concerned with an efficient structure of an organization when the efficiency is measured by the ratio between a production (primary activity) capacity and the number of the organization members. The production capacity is measured by the number of the operators who are directly involved in the production activity.

The results of this chapter are restatements of Ref. [14] although the proofs are modified.

4.1 Hierarchical Skeleton of Organization

Fig. 4.1 illustrates a typical scheme of the organization considered in this chapter.

As the figure indicates, an organization is assumed structured as a directed tree whose nodes represent the members and whose arcs are reporting relations among them. The top node denotes the president, p. A leaf node represents an operator, who carries out a real task or the production activity of the organization.

Since no structure is assumed for an organization member, every member is supposed to have the same capability. The production capacity of an organization is, therefore, assumed simply proportional to the number of operators.

We will start from formalization of Fig. 4.1 following Ref. [14].

Definition 4.1 Organization skeleton

Let

N = the set of organization members

where n= |N|. Then an *organization skeleton* is represented by the following structure:

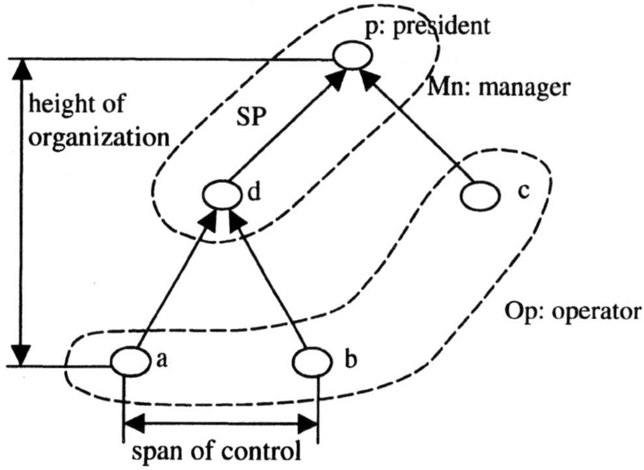

Figure 4.1 An organization structure

organization skeleton = <N,SP,p>

where

$p \in N$

is the president and SP is a function

$SP: N \rightarrow N$

such that SP satisfies the following conditions:

(i) Irreflexive: $i \neq p \rightarrow i \neq SP(i)$;

(ii) Connectivity to p: $p=SP(p)$ and $(\exists\, i \neq p)(p=SP(i))$;

(iii) Acyclic: $j_1=SP(i), j_2=SP(j_1),..,j_n=SP(j_{n-1}), i \neq p \rightarrow i \neq SP(j_n)$.

SP will be called *supervisor function.* When y=SP(x), y is a supervisor of x which is unique because SP is defined as a function.

The condition (i) implies that any member except the president cannot be his supervisor. The condition (ii) says that the supervisor of the president is the president himself and that he has at least one subordinate. The condition

(iii) requires that the supervising relation does not have a loop or anybody except the president can be his own supervisor neither in an indirect way nor in a direct way.

The acyclic property implies that

Asymmetric: $j=SP(i),\ i \neq p \rightarrow i \neq SP(j)$.

It will be shown later that the conditions (i), (ii) and (iii) imply that the supervising relation generated by SP is a tree structure whose top node is p.

Using the function SP we can formalize classical concepts of an organization.

Definition 4.2 Class of managers and class of operators
Let

$$Mn = \{i \mid (\exists j)(i=SP(j))\}$$

and

$$Op = N - Mn.$$

Mn and Op are called set of *managers* and *set of operators*, respectively.

A manager is defined as a member who has at least one subordinate. In particular, the president is a manager because of the connectivity condition. A member who is not a manager is an operator.

This chapter assumes that the real primary activity of an organization is carried out only by operators although in reality there may be many "playing managers".

For the sake of notational convenience we introduce the following notation.

Definition 4.3 Reporting path
Let

(i) $i \rightarrow j$ if and only if $j=SP(i)$

and

(ii) $i \rightarrow \rightarrow j$ if and only if $(\exists\, i_1,..,i_m)(i=i_1 \rightarrow \cdots \rightarrow i_m \rightarrow j)$

where $i_{k+1}=SP(i_k)$.

The arrow $i \to \to j$ is called *reporting path* from the member i to the member j.

Definition 4.4 Tree structure on organization members
The following structure is called *tree structure* on N with the top node n_0 :

$$Tree=<N,R,n_0>$$

where

$$R \subset N \times N$$

such that

(i) $n_0 \in N$

(ii) $(\forall\, i \neq n_0)(\exists\, !\, j)(j \neq i \text{ and } (i,j) \in R)$

(iii) $(n_0,i) \in R \leftrightarrow i=n_0 \text{ and } (\exists\, i)(i \neq n_0 \text{ and } (i,n_0) \in R)$

(iv) $(i,j_1) \in R, (j_1,j_2) \in R,..,(j_{n-1},j_n) \in R, i \neq n_0 \to (j_n,i) \notin R.$

Then we have

Lemma 4.1
For <N,SP,p> let Org $\subset N \times N$ be such that

$$(i,j) \in Org \leftrightarrow j=SP(i).$$

Then, <N,Org,p> is a tree structure where p is the unique top node. Furthermore, if Org is extended to Org' $\subset N \times N$ as

$$(i,j) \in Org' \leftrightarrow j=i \text{ or } i \to \to j,$$

Org' is a partially ordered relation.
The proof is given in Appendix.

Lemma 4.1 says that the organization structure given by Def. 4.1 is a tree structure.

Definition 4.5 Chain of command

If $i=i_1 \to \cdots \to p$, the reporting path $i_1 \to \cdots \to p$ is called *chain of command.*

Lemma 4.2

There is a function C:N $\to$ {chain of command on N} such that

$$C(i)=(i=i_1 \to\to p).$$

The proof is given in Appendix.

Lemma 4.2 implies that there is a chain of command for any i in N and it is unique.

The height of an organization is defined by the function C. Let length(C(i)) be the length of the chain C(i), i.e.,

$$\text{if } C(i)=(i \to p), \text{length}(C(i))=1$$

and

$$\text{if } C(i)=(i \to j \to \cdots \to p), \text{length}(C(i))=\text{length}(C(j))+1.$$

Definition 4.6 Height of an organization

Let

$$H=\max \{\text{length}(C(i)) \mid i \in N \}.$$

Then, H is called *height of an organization.*

Definition 4.7 Rank of organization member

Let

$$\text{Rank}:N \to I$$

be a function which is defined recursively. Let H be the height of an organization.

(i) Rank(p)=H.

(ii) if Rank(i)=m and i=SP(j), then Rank(j)=m−1.

The span of control can be defined by the function SP in a straight forward way.

Definition 4.8 Span of control
Let $S:N \rightarrow \wp(N)$ be such that

$$S(i)=\{j|i=SP(j)\}-\{p\}.$$

Then $|S(i)|$ is called *span of control* of i.

S(i) is the set of members whom i supervises directly. Notice that p is excluded from S(p).

4.2 Taxonomy of Organization

Using the above concepts we can define specialized structures of an organization [14].

Definition 4.9 Balanced organization
An organization is called *balanced* if and only if its skeleton satisfies

$$(\exists\, i_0 \in Op)(\forall\, i)(i \in Op \rightarrow \text{length}(C(i))=\text{length}(C(i_0))),$$

that is, an organization is balanced if and only if lengths of chain of commands from operators are equal to each other.

Definition 4.10 Constant span of control organization
An organization is said to have a *constant span of control* if and only if its skeleton satisfies

$$(\exists\, i_0 \in Mn)(\forall\, i)(i \in Mn \rightarrow |S(i)|=|S(i_0)|),$$

that is, an organization is of a constant span of control if and only if each manager has the same number of direct subordinates where p is not included in the set of subordinates of p.

Definition 4.11 Regular organization
An organization skeleton is called *regular* if and only if it is balanced and of constant span of control.

Fig. 4.2 illustrates concepts of the specialized structures.

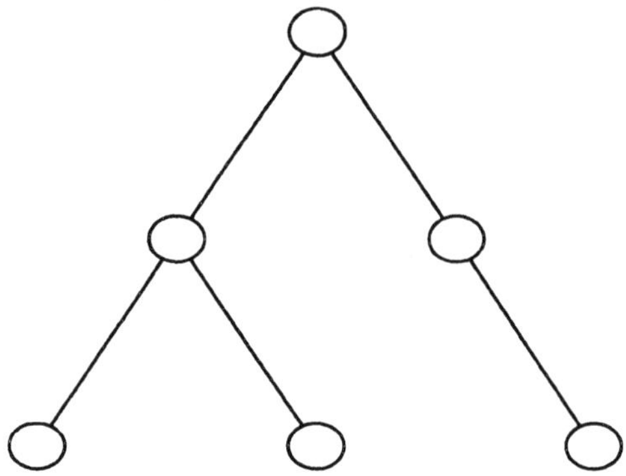

Balanced organization

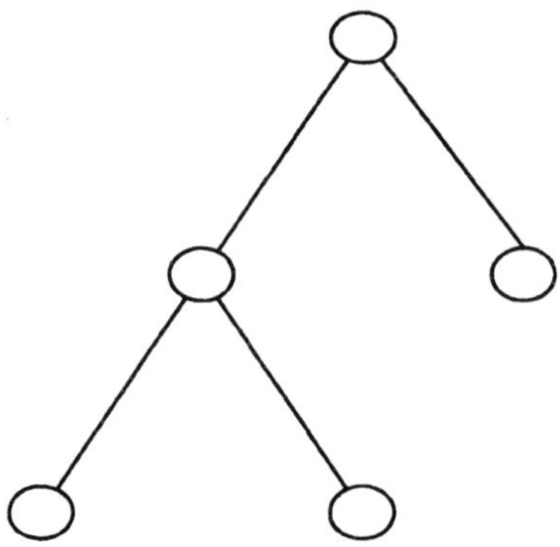

Constant span of control organization

Figure 4.2 Taxonomy of organization

They are ideal types of organization structure.

4.3 Properties of Organization Chart

For the specialized structures introduced above we can prove the following facts [14].

Proposition 4.1

Suppose an organization has the constant span of control property whose span is s. Suppose Op is a given operator set. Then, |Mn| is determined by

$$|Mn|=(|Op|-1)/(s-1).$$

The proof is given in Appendix.

This relation is important because it holds regardless of the structure of an organization as long as the structure is of a constant span of control. Then,

Corollary 4.1
Under the condition of Prop. 4.1 the following relation holds:

$$|N|=|Mn|+|Op|=(s|Op|-1)/(s-1).$$

Then, we have the following fact.

Theorem 4.1
Under the condition of Prop. 4.1 the productivity measured by |Op|/|N| decreases monotonously regardless of its structure as the number of members of an organization increases.

The proof is given in Appendix.

Although the proofs of these results are not difficult, the results are not necessarily self-evident and furthermore they are useful for rough estimation of basic properties of an organization structure.

The next result also provides a basis to an argument of efficiency of an organization structure.

Let H be the height of a balanced organization.

Let s_r (r=1,..,H) be the span of control of the r-th layer (rank) of the organization where every member on the same rank is assumed to have the same span of control.

Let $\underline{s}$ be:

$$\underline{s}= (\Sigma_{r=1}^{H} s_r)/H .$$

$\underline{s}$ is the average span of control of the organization.

Notice that in the present case the number of the operators, |Op|, of the organization is given by:

$|Op| = \Pi_{r=1}^{H} s_r.$

Then,

Proposition 4.2

Suppose the height of a balanced organization H is fixed. Suppose an average span of control $\underline{s}$ is given. Then the number of operators (=$|Op|$) is maximized when all span of control are equal.

The proof is given in Appendix.

Lemma 4.3

Let H be the height of an organization. Suppose the organization is of constant span of control whose span is s. Then, the following relation holds:

$$|N| \leqq 1+s+\cdots+s^H.$$

The proof is given in Appendix.

Corollary4.2

For a regular organization the following relation holds:

$$|N|=1+s+\cdots+s^H$$

where s is the span of control and H is the height of an organization.

Proposition 4.3

Suppose an organization skeleton is of a constant span. Then, it satisfies the following properties:

(i) For a given s and H $|Op|$ is maximized for a regular organization.

(ii) For a given s and N_{op} (integer), the height H of an organization is minimized for a regular organization where $N_{op}=|Op|$ is required.

(iii) For a given N_{op} (integer) and H, the span of control s is minimized for a regular organization where $N_{op}= |Op|$ is required.

The proof is given in Appendix.

APPENDIX OF CHAPTER 4

Proof of Lemma 4.1

The fact that <N,Org,p> is a tree structure is clear from Def. 4.1. Hence, let us consider the second part. We have to show that Org' satisfies the axioms of the partially ordered relation.

(i) Reflexive:

Let $i \in N$ be arbitrary. Then $(i,i) \in Org'$ is true by definition.

(ii) Anti-symmetric:

Suppose $(i,j) \in Org'$ and $(j,i) \in Org'$. Suppose $i \to \cdots \to j$ and $j \to \cdots \to k \to i$. Then, we have $i \to \cdots \to j \to \cdots \to k \to i$. Then, (iii) of Def. 4.1 implies that i=p, that is, j=p. Therefore, i=j. In the other cases we have i=j immediately.

(iii) Transitive:

Suppose $(i,j) \in Org'$ and $(j,k) \in Org'$. Suppose $i \to \cdots \to j$ and $j \to \cdots \to k$. Then, we have $i \to \cdots \to j \to \cdots \to k$ or $(i,k) \in Org'$. In the other cases we have $(i,k) \in Org'$ easily.

Q.E.D.

Proof of Lemma 4.2

We have to prove two facts.

a. We will show that for any $i \in N$ there is a relation $i \to\to p$. Let parents(i)= $\{j_n | j_n = SP^n(i)$ for $n=0,1,2,..$ where $j_0=1\}$. Since parents(i) $\subset N$ and N is finite, there must be j_k and j_l such that $j_k = j_l$. Let us consider the sequence $j_k \to j_{k+1} \to \cdots \to j_l$ where k<l is assumed. Then, (iii) of Def. 4.1 implies that j_l=p. Therefore, a relation $(i \to \to p=j_l)$ exists.

b. Next we will show a path from i to p is unique. Suppose there are two sequences $i \to j_1 \to \cdots \to p$ and $i \to j_1' \to \cdots \to p$. Let k be the smallest integer such that $j_l=j_l'$ for $l \leqq k$ and $j_{k+1} \neq j_{k+1}'$. Then, $SP(j_k)$ has two values which contradicts the functionality of SP.

Q.E.D.

Proof of Proposition 4.1

Suppose every manager checks a mark on his direct subordinates. Then, every member of the organization is checked once and only once except p. p is not checked because he is not considered as a subordinate of his. Then, the following relation should hold:

$$|Mn|s=(|Mn|+|Op|-1).$$

Consequently, we have

$|Mn|=(|Op|-1)/(s-1).$

Q.E.D.

Proof of Theorem 4.1
Since

$$|Op|=(|N|(s-1)+1)/s,$$

we have

$$|Op|/|N|=(s-1)/s+1/(|N|s).$$

Therefore, $|Op|/|N|$ decreases monotonously as $|N|$ increases.
Q.E.D.

Proof of Proposition 4.2
We use the following inequality:

$$(\Pi_{i=1}^{H} s_r)^{1/H} \leqq (\Sigma_{r=1}^{H} s_r)/H.$$

This inequality comes from the following simple fact. Notice that the log function is a concave function (Refer to Appendix II). Therefore, we have the following inequality:

$$\log(a_1x_1+\cdots+a_nx_n) \geqq a_1\log(x_1)+\cdots+a_n\log(x_n)$$

where $a_i \geqq 0$, $x_i > 0$ and $a_1+\cdots+a_n=1$. Let $a_i=1/n$. Then, from the above inequality we have

$$\log((x_1+\cdots+x_n)/n) \geqq (\log(x_1)+\cdots+\log(x_n))/n$$

or

$$\log((x_1+\cdots+x_n)/n) \geqq \log(x_1*\cdots*x_n)/n$$

or

$$\log((x_1+\cdots+x_n)/n) \geqq \log((x_1*\cdots*x_n)^{1/n})$$

or

$$(x_1+\cdots+x_n)/n \geqq (x_1*\cdots*x_n)^{1/n}.$$

Then

$$|Op| \leqq (\underline{s})^H$$

and the equality holds if and only if $s_r = \underline{s}$ for all r.

Q.E.D.

Proof of Lemma 4.3

We will prove by the mathematical induction.

(i) Suppose H=1 where the skeleton consists of the president and his direct subordinates. Then,

$$|N|=1+s \leqq 1+s^1.$$

(ii) Suppose the lemma is true for H $\leqq$ r. Let H=r+1. Let N_i (i=1,..,s) be the number of the members under the i-th subordinate (including himself) of the president. Then, the above hypothesis implies

$$|N_i| \leqq 1+\cdots+s^r.$$

Since

$$|N|=|N_1|+\cdots+|N_s|+1,$$

we have

$$|N| \leqq (1+\cdots+s^r)*s+1 =1+\cdots+s^{r+1}.$$

Q.E.D.

Proof of Proposition 4.3

Let us consider (i). Let n_r be the number of the positions of the rank r of an organization. Then, if the organization is balanced, $n_r=s^{H-r}$. As a matter of fact, $n_H=1, n_{H-1}=s, n_{H-2}=s^2$,..... Then, Lemma 4.3 and Cor. 4.2 say that for fixed H and s N is maximized at a balanced organization.

On the other hand, Prop. 4.1 says

$$|Op|=1/s + (s-1)|N|/s.$$

Therefore, |Op| is maximized at the balanced organization.

Let us consider (ii). Suppose there exists a regular organization whose height is H^* and which satisfies s and $N_{op}=|Op|$ of the proposition. Let us assume that there is an organization architecture A whose height is $H' < H^*$ and which satisfies s and the number of operators=$N_{op.}$ In general, let $Op(s,H) = s^H$ be the number of Op for a regular structure whose height is H and whose span of control is s. Then, since the number of operators of A is N_{op}, (i) implies

$$Op(s,H') \geqq N_{op}.$$

Since, $N_{op}=Op(s,H^*)$, we have

$$Op(s,H') \geqq Op(s,H^*).$$

Since $H' < H^*$ implies that

$$Op(s,H') < Op(s,H^*)\ ,$$

we have a contradiction.

Let us consider (iii). Let s be the span of control of a regular organization such that $N_{op}=Op(s,H)$. Suppose there is $s'< s$ and a structure A such that A satisfies the number of operators=N_{op} and H. Then, $s'<s$ implies

$$N_{op} = Op(s,H) > Op(s',H).$$

Furthermore, since A satisfies N_{op} and H, (i) implies

$$Op(s',H) \geqq N_{op},$$

which contradicts the above relation.

Q.E.D.

CHAPTER 5 OPERATIONAL LEVEL MANAGEMENT

REGULATION AND OPTIMIZATION

Since hierarchy is one of the most fundamental characteristics of our organization model or, in fact, in general of any real organization, Chapter 4 explored organization properties focusing only on the hierarchical aspect of an organization as a starting point of our organization theory. The result is a construction of a formal theory of an organization chart. The other fundamental character of an organization, its goal seeking activity, will be investigated in this and subsequent chapters. This book assumes that an organization and its components are goal seeking systems except in Chapter 11.

Armed with the presentation of a goal seeking system concept in MGST introduced in Chapter 2 and Chapter 3, we will now investigate, step by step, the management behavior of each level along the hierarchy, from the bottom to the top, so that a comprehensive view of the cybernetic approach to an organization can be revealed in the mathematical way. This chapter is concerned with the bottom, operational, level.

The basic functions of the operational level management are its regulation and optimization activities. Regulation and optimization are main topic of the control engineering and the management science. The results developed in these fields are based on quantitative models, in contrast to the treatments in this book which on the whole is in the context of MGST which, in turn, uses only non-numerical mathematical apparatus. We treat structural problems to which numerical results in control and management science have little to contribute. This in no way is to question the usefulness of numerical treatments. Indeed, numerical and non-numerical methods are complementary and address problems of the same "entity", i.e., organization from different angles. This chapter will investigate a formal behavior of the operational level management by using concepts common to both domains.

Section 5.1 will discuss the regulation activity of the operational level. The regulation is implemented by a feedback mechanism. The objective of a feedback mechanism is to bring the state of a regulated object to a desirable state—also called a set point—which is specified by the optimization activity in coordinatio—and to maintain the system at the desirable state against a disturbance input. The latter function is a topic of the stability theory.

In Section 5.1 we assume causality and stationarity properties for the dynamics of the process of the operational level and derive a system's state space representation. In Section 5.2 a qualitative stability theory will be presented for the state space representation of a dynamical system. The mathematical results are presented without proof because their proofs require a more extensive background. The proofs are given in Ref. [2].

In Section 5.3 the optimizing activity of the operational level is discussed. The regulated behavior of the process is the target of the optimizing management of the operational level. The set point of the regulation is specified by the optimization, as mentioned above. It must be noticed that optimization is a valid management activity of the operational level because the level is well protected from the external input by the upper level managements.

5.1 Regulation of Operational Level

a. Formulation and Representation of the Overall Process

Since the subject of the operational management is a technological system which can be described in a numerical way, the results in control and management science have received wide application. This chapter addresses the problems traditionally in the domain of control and management science but in the abstract framework of MGST.

We start from an image of an overall process given by Fig. 5.1.

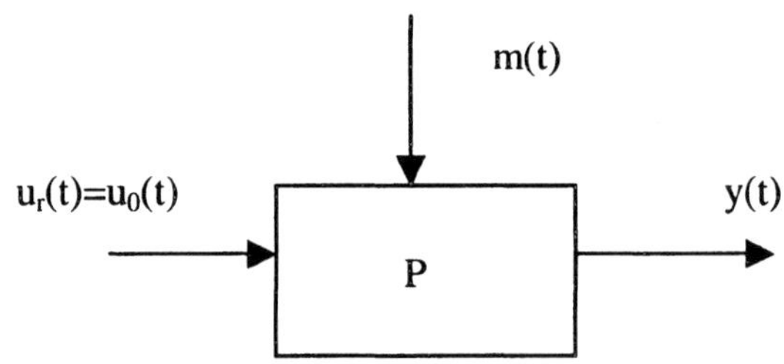

Figure5.1 Model of Overall Process

The process has two inputs: manipulating variable input *m* and resource input u_r. By defining $x=(m,u_r)\in X$, the process P will be treated as an input output system on X and Y.

First we introduce two assumptions for the process P.

Assumption 1

The overall process P is a closed system because the resource input is assumed either known or controllable.

Assumption 2

The overall process is a stationary dynamic system over a regulation time span (scale) T_{reg}. In the numerical domain it is exemplified as a discrete system.

This book assumes three types of time scales: adaptation time scale T_{adapt}, coordination time scale T_{coord}, and regulation time scale T_{reg}. Management activities, adaptation, coordination and regulation are performed with respect to the corresponding time scale in mind. The lengths of the time scales are decreasing from T_{adapt} to T_{reg}.

The second assumption asserts that the dynamic behavior of the operational level in the numerical domain is represented by a difference equation, which can cover a wider range of models than a differential equation. For instance, an automaton model can be described by a difference equation.

Let u_0 be a known input of the management resource. Then

$$X=M \times \{u_0\}$$

and

$$P: X \rightarrow Y .$$

Since an organization is a dynamic system, it is necessary to represent its variables $x \in X$ and $y \in Y$ as time functions. Let us introduce the concept of a time function from MGST.

Definition 5.1 Time set

Let the following set

$$T=\{0,1,2,\ldots\}$$

be called *time set* for a discrete dynamic system.

Definition 5.2 Input alphabet
Let

A = set of values which elements of X take at each time.

In other words, $x(t) \in A$ holds for $x \in X$ and $t \in T$. A is called *input alphabet.*

Definition 5.3 Output alphabet
Let

B= set of values which elements of Y take at each time.

Then, B is called *output alphabet.*

Definition 5.4 Time function
Time functions x and y and their sets X and Y, respectively, are defined as:

$$X \subset \{x|x:T \to A\}$$

and

$$Y \subset \{y|y:T \to B\}.$$

A process model is represented as a function in Chapter 2 and Chapter 3. However, when P is to represent a dynamics of the process, in which its output depends on the initial state of the system as well as on the input x, P is in general a relation over its inputs and outputs. An output value cannot be determined in such a case uniquely by an input alone. In the subsequent discussions of this chapter we use the following as a general definition of P :

Definition 5.5 General representation of process
Let a *general representation* **of the process** be the following input and output relation:

$$P \subset X \times Y.$$

In order to describe the dynamic behavior of an organization let the restriction operations on sets and functions be introduced as follows:

$T^t=[0,..,t)$,

$\underline{T}^t=[0,..,t]$,

$x^t=x|[0,..,t)$,

$x_{ts}=x|[t,..,s)$,

$x_t=x|[t,t+1,..)$.

Let a shift operation σ^s for a function be:

$\sigma^s(x_{tu})(i)=x(i-s)$

where $t+s \leqq i < u+s$. That is, σ^s shifts a time function along the time coordinate by s. The same definitions will be applied to $y \in Y$ and other time functions.

Fig. 5.2 illustrates the concepts of restricted functions and the shift operation.

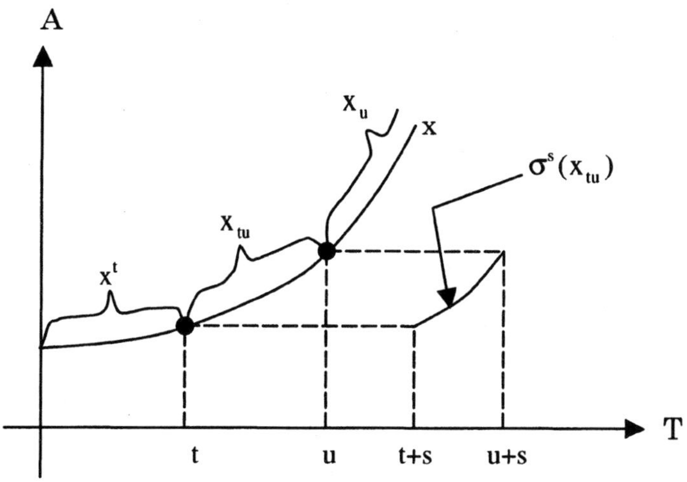

Figure 5.2 Restrictions and shift operation

The restriction operation and the shift operation are extended to sets. For instance,

$$X^t = X|T^t = \{x^t | x \in X\}.$$

Using the above definitions we can have formal representations of the assumptions about the process.

Definition 5.6 Stationary process

$P \subset X \times Y$ is called *stationary* if and only if the following holds:

$$(\forall t)(\sigma^{-t}(P|T_t) \subset P).$$

Since P represents a physical (technological) system, it should satisfy a causality property. The following is a formal definition of the (non-anticipation type) causality.

Definition 5.7 Causal process

$P \subset X \times Y$ is called *causal* if and only if the following holds:

$$(\forall t)(\forall x,x')(x^t = x'^t \rightarrow P(x)|\underline{T}^t = P(x')|\underline{T}^t)$$

where $P(x) = \{y | (x,y) \in P\}$.

Then a main result can be presented as follows [2]:

Theorem 5.1 (Discrete system case)

Suppose P satisfies Def. 5.6 and Def. 5.7. Then, P has a time invariant state space representation. That is, there exist a set C and two functions $\phi: C \times A \rightarrow C$ *and* $\lambda: C \rightarrow B$ *and for the set* $Z = \{z | z: T \rightarrow C\}$ *the following relations hold:*

$$(i) \qquad z(n+1) = \phi(z(n), x(n)) \; (n=0,1,\ldots)$$

$$(ii) \qquad (x,y) \in P \leftrightarrow (\exists c)(\forall t)(y(t) = \lambda(\bar{\phi}(c, x^t)))$$

where $\bar{\phi}$ *is recursively defined as follows:*

$$\bar{\phi}^0(c, x^0) = c,$$

$$\phi^{t+1}(c,x^{t+1})=\phi(\phi^t(c,x^t),x(t)).$$

C, φ, λ, Z are called *state space, state transition function, output function* and *trajectory set of the state,* respectively. The state space representation is the basis for the discussion of a dynamic system.

b. Dynamic Optimization Formulation of Regulation Problem

The operational management activity actually consists of two levels, regulation and optimization. The optimization is performed at the coordination planning stage where negotiation takes place to resolve the conflicts on the operational level. The regulation is performed to secure the execution of the coordination plan.

The regulation activity is specified as a result of dynamic optimization. This section will introduce a dynamic optimization problem for the regulation activity.

Suppose an optimum outcome $y^* \in Y$ is prescribed by the optimization activity.

Let $z^* \in Z$ be

$$\lambda(z^*)=y^*.$$

z^* represents a desired trajectory of state. The objective of the regulation activity is to ensure the desired state trajectory z^*.

Let us introduce the third assumption.

Assumption 3:

The change of z^* in time is taking place at a slow rate compared with that of the dynamics of P.

Let

T_{coord}=time span of coordination,

T_{reg}=time span of regulation.

A coordination plan is specified for a given time interval which is denoted by T_{coord}. For instance, T_{coord} may be 3 months. On the other hand, the time span T_{reg} for which a dynamic optimization is performed is much shorter. T_{reg} may be a week, a day, or even less.

Let

$L = T_{coord}/T_{reg}$ (integer).

Let

$c^h_l = z^*(l * T_{reg}) \in C$ ($l = 0,1,2,\ldots$).

Then, Assumption 3 indicates

$z^*(t) \approx c^h_l$ for $l * T_{reg} \leqq t < (l+1) * T_{reg}$.

c^h_l will be called a *set point* for a dynamic optimization. The regulation activity is to maintain the state of P at c^h_l during the period from $T_{reg} * l$ to $T_{reg} * (l+1)$.

The subsequent discussions will focus on the period from 0 to T_{reg}. For the notational convenience c^h_0 will be denoted by c^h. Fig. 5.3 illustrates relations of the above definitions.

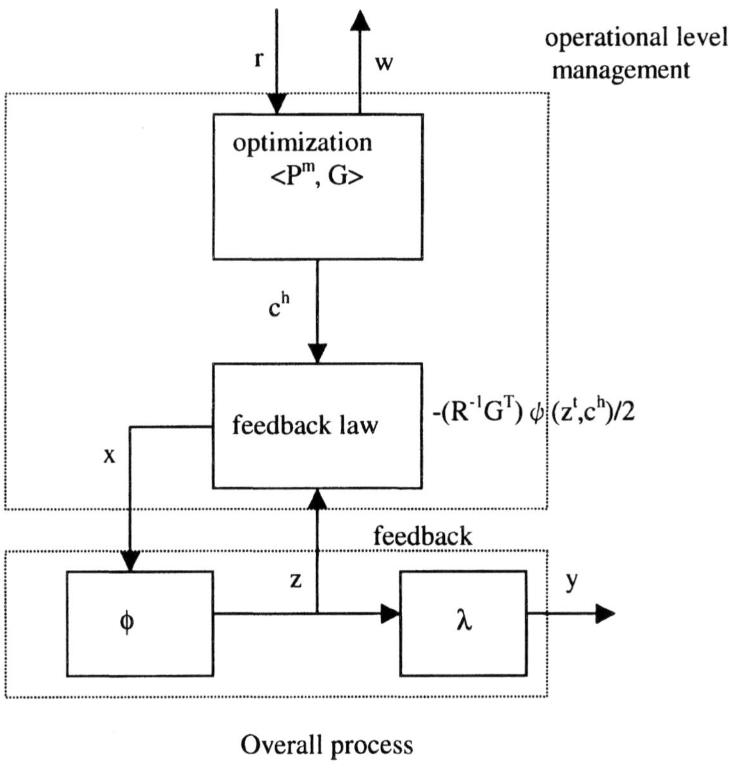

Figure 5.3 Regulation mechanism of operational level management

A typical example of a dynamic optimization for the regulation is the following LQP(linear quadratic problem).

$$z(n+1)=Fz(n)+Gx(n),$$

$$y(n)=Hz(n),$$

$$G_{reg}(x,c^h,z)=\Sigma\,[(z(n)-c^h)^TQ(z(n)-c^h)+x(n)^TRx(n)],$$

where F, G, H, Q and R are constant matrices on reals. In particular, Q and R are symmetric and positive, i.e.,

$$z(n)^TQz(n)\geqq 0,$$

$$x(n)^TRx(n)\geqq 0.$$

G_{reg} is required to be minimized.

The term $x(n)^TRx(n)$ represent a production driven aspect of the regulation goal.

Although the LQP may be too simple to describe real situations, this chapter will use it to explore the basic conceptual mechanism of the operational level management.

The dynamic optimization problem is, then, in general given as:

$$LQP=\langle X,C,Z,P,G_{reg},c^h\rangle$$

where

$$P:C\times X\rightarrow Z,$$

$$Y=Z,$$

$$G_{reg}:X\times C\times Z\rightarrow Re$$

and

$c^h \in C$.

An optimal solution x^* is one which minimizes G_{reg} for a given c^h.

c. Regulator Problem of Operational Level

Regulation of the operational level is a feedback form implementation of a solution x^* of the dynamic optimization problem. The feedback form is given by the following feedback law: For any t

$$x^*(t)=FB(z^t).$$

That is, $x^*(t)$, the value of the optimal solution at time t is given by a function of z^t ,the past trajectory of the state. A typical example of FB in control engineering is so called PID controller which is given by

$$x^*(t) = \alpha z(t) + \beta \int_0^t z(t)dt + \gamma(dz/dt)$$

where α, β and γ are constant parameters. It should be noticed that $x^*(t)$ is specified as a function of z^t in a PID controller.

The problem to find an FB is called *a regulator problem.* Let us investigate the regulator problem using the LQP.

Although our model is discrete, we will use the formulation for a continuous system in order to use the result of Appendix III directly. Then, a dynamic optimization problem is given by:

$$dz/dt=Fz+Gx,$$

$$G_{reg}(x,c^h,z)= \int_0^{T_{reg}} (z-c^h)^T Q(z-c^h)+x^T Rx\,dt$$

It is not easy to solve a dynamic optimization problem. Typical techniques are dynamic programming and the maximum principle. This book uses the maximum principle which is explained in Appendix III. Then, the maximum principle says that an optimum solution x^* is given by minimization of the following Hamiltonian:

$$H=(z-c^h)^T Q(z-c^h)+x^T Rx+p^T(Fz+Gx)$$

where p is called adjoint variable which corresponds to a Lagrangean multiplier. That is, x^* is given by

$$H_x=0$$

or

$$2Rx^*+G^Tp=0$$

or

$$x^*=-(R^{-1}G^Tp)/2$$

where R is assumed to have the inverse.

Furthermore, p and z are given by:

$$dp/dt=-H_z=-(2q(z-c^h)+FTp) \qquad \text{(A)}$$

and

$$dz/dt=Fz+G(-(R^{-1}G^Tp)/2),$$

where the following boundary value conditions are to be satisfied:

$$z(0)=\text{given initial condition},$$

$$p(T_{reg})=0.$$

(A) shows that if z is found, p is given by the following form:

$$p(t)=\phi(z^t,c^h).$$

Then, FB is given by

$$x^*(t)=-(R^{-1}G^T\phi(z^t,c^h))/2.$$

This is a solution of the regulator problem. Fig. 5.3 shows the derived regulation mechanism of the operational level management.

5.2 Stability of Operational Level

In general, let FB be represented as:

$$FB(c^h):Z\rightarrow A^T$$

where $x(t)=FB(c^h)(z)(t)$ and if $z^t=z'^t$. Then $FB(c^h)(z)|T^t= FB(c^h)(z')|T^t$. FB is parameterized by the set point $c^h \in C$.

Substituting the feedback relation into the state space representation of the process, the behavior of the process under the regulation is given by:

$$z(n+1)=\phi(z(n),FB(c^h)(z)(n)).$$

When the above difference equation is solved for an initial condition

$$z(0)=c\ ,$$

the solution can be represented by a function

$$\Pi:C \times T \to C,$$

or

$$z(t)=\Pi\ (c,t).$$

Π is called dynamical system representation.

In general

Definition 5.8 Dynamical system representation of process

$\Pi:C \times T \to C$,which satisfies the following properties:

$$\Pi(c,0)=c\ ,$$

$$\Pi(c,t+s)=\Pi(\Pi(c,t),s),$$

is called *dynamical system representation* [2]. Π presents the dynamic behavior of the process under the feedback regulation.

The objective of the regulation is to maintain the state at c^h against any disturbances. If the objective is realized, c^h is called stable. Let us consider how a qualitative argument can be developed for the stability of the process using Π.

Let us define a singular point $c^* \in C$ which is called equilibrium state (point).

Definition 5.9 Equilibrium state

Let $c^* \in C$ satisfy the following relation:

$$(\forall n)(\Pi(c^*,n)=c^*).$$

c^* is called an *equilibrium state*.

Suppose the state of the process is equal to c^*. Then, if the state is displaced from c^* due to a disturbance from the environment, a good regulation is required to bring the state to c^* by the feedback function. We will say the equilibrium point c^* is stable if this desired property holds.

Since c^h is the set point of the feedback regulation, let us say that the regulation is satisfactory when $c^*= c^h$ and c^* is stable.

In order to formalize the stability concept a displacement concept or a neighborhood concept should be introduced. Mathematically the neighborhood concept is specified by topology.

Let

$$\Theta_c=\{C'_i|C'_i \subset C \wedge i \in I\}$$

be a class of "open sets" over the state space C. It should be noticed that the class Θ_c is allowed not to satisfy the usual definition of the topology. It can be an extended topology. Of course it can satisfy the usual definition of topology as well.

The class Θ_c is usually specified to express some neighborhood concept related with a behavior of an organization. The following is an example: Suppose a desirability criterion of the state of the process is specified by a multi-objective function

$$S:C \to \{0,1\}^P.$$

Then, S(c) is a P dimensional vector and if the i-th element of the vector is 1, it indicates that the i-th aspect of the criteria is satisfactory for c. Otherwise, it is not satisfactory. The set $\{0,1\}^P$ can be made a partially ordered set introducing an order relation $\leqq_s$ in the usual way, i.e.,

$$v \leqq_s v' \leftrightarrow (\forall i)(v(i) \leqq v'(i))$$

where v(i) is the i-th element of $v \in \{0,1\}^P$ and the relation $0 \leqq 1$ is assumed.

Let

$$\rho : C \times C \to \{0,1\}^P$$

be such that

$$\rho(c,c')(i) = \begin{cases} 0 \text{ if } S(c)(i) = S(c')(i) \\ 1 \text{ if } S(c)(i) \neq S(c')(i) \end{cases}$$

ρ is called a *pseudo distance function*. Then, an extension of the usual topology is defined as follows:

For $c \in C$ and $v \in \{0,1\}^P$ let

$$S_v(c)=\{c' \mid \rho(c,c') \leqq_s v\}.$$

Let

$$C' \in \Theta_c \leftrightarrow (\forall c \in C')(\exists v)(S_v(c) \subset C').$$

When a "topology" Θ_c is given, the stability concept is formalized in the following way:

Definition 5.10 Lyapunov like stability

An equilibrium point c^* is called *stable* if and only if the following condition is satisfied:

$$(\forall C_i \ni c^*)(\exists C_j \ni c^*)(\forall c)(\forall n)\ (c \in C_j \rightarrow \Pi(c,n) \in C_i).$$

The above definition of stability is an abstraction of the usual definition of Lyapunov stability.

We have the following result about the Lyapunov like stability.

Definition 5.11 Lyapunov type function

Let c^* be an equilibrium point of Π. Let N be a partially ordered set and let $N^+ \subset N$. Suppose there is a function $L:C \rightarrow N$ which satisfies the following conditions:

(i) $(\forall c)(\forall n)(L(\Pi(c,n)) \leqq L(c))$

(ii) $(\forall C_i)(\exists r \in N^+)(\forall c)(c^* \in C_i \wedge L(c) \leqq r \rightarrow c \in C_i)$

(iii) $(\forall r \in N^+)(\exists C_j)(\forall c)(c^* \in C_j \wedge c \in C_j \rightarrow L(c) \leqq r)$.

Then, L is called *Lyapunov type function* for c^*.

A Lyapunov type function is another abstraction of the usual definition of Lyapunov function.

Under the above general definitions we can prove the following fact [2].

Theorem 5.2

An equilibrium point $c^* \in C$ *satisfies the Lyapunov like stability if and only if there exists a Lyapunov type function for* c^*.

If a system is described by a linear time invariant difference (differential) equation, stability can be checked directly by the representation itself. But in the general case, stability is examined by finding an appropriate Lyapunov function. The above theorem provides a foundation for that approach.

5.3 Departmentalized Operational Units

Let us extend the above general results to the situation of the operational level whose process consists of interacted sub-processes. Fig. 5.4 illustrates the situation.

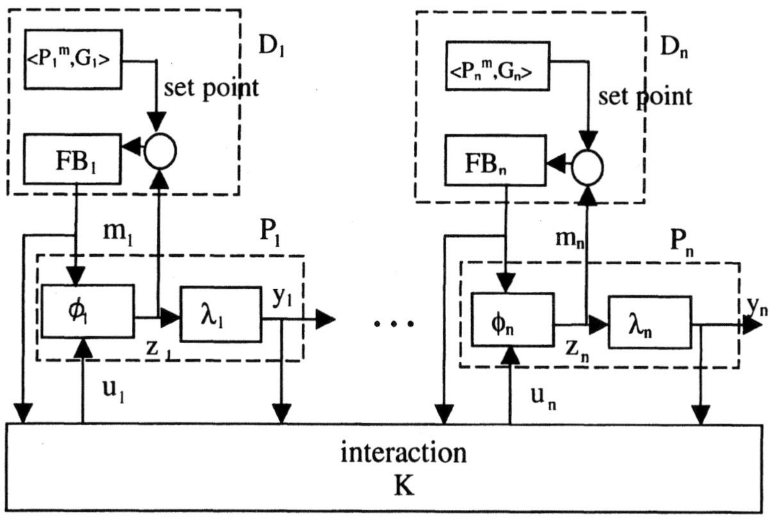

Figure 5.4 Operational level management

Let the i-th subprocess be

$P_i \subset M_i \times U_i \times Y_i$

where $m_i \in M_i$ and $u_i \in U_i$ are a manipulating variable and an interaction input, respectively.

Let

$$M = M_1 \times \cdots \times M_n,$$

$$Y = Y_1 \times \cdots \times Y_n,$$

and

$$U = U_1 \times \cdots \times U_n.$$

Let

$$K_i: M \times Y \to U_i$$

or

$$u_i(k) = K_i(m(k), y(k)).$$

K_i is a static function.

Let

$$\underline{K}: M \times Y \to U$$

be

$$\underline{K}(m,y) = (K_1(m_1,y_1),\ldots,K_n(m_n,y_n)).$$

Then, the above general consideration implies the following state space representation of P_i :

$$\phi_i: C_i \times A_i^m \times A_i^u \to C_i$$

and

$$\lambda_i: C_i \to B_i$$

where A_i^m and A_i^u are the alphabets of M_i and U_i, respectively. C_i is the state space of P_i.

Let $\phi: C \times A^m \times A^u \to C$ be

$$\phi(c,a^m,a^u)=(\phi_1(c_1,a_1{}^m,a_1{}^u),\ldots,\phi_n(c_n,a_n{}^m,a_n{}^u))$$

where $a^m \in A^m=A_1{}^m \times \cdots \times A_n{}^m$ and $a^u \in A^u=A_1{}^u \times \cdots \times A_n{}^u$.

The state space representation implies the following functional representation of P_i :

$$P_i:C_i \times M_i \times U_i \to Y_i.$$

Let $\underline{P}:C \times M \times U \to Y$ be

$$\underline{P}(c,m,u)=(P_1(c_1,m_1,u_1),\ldots,P_n(c_n,m_n,u_n)).$$

Then, a functional form representation of the operational level process

$$P:C \times M \to Y$$

is derived from $(\underline{P},\underline{K})$ by eliminating u.

Let $K_i':C \times M \to U_i$ be

$$K_i'(c,m(k))=K_i(m(k),P(c,m)(k)).$$

Let

$$\underline{K}'(z(k),m(k))=(K_1'(z_1(k),m_1(k)),\ldots,K_n'(z_n(k),m_n(k))).$$

Let the feedback of the i-th unit $FB_i(c^h{}_i):C_i \to A_i{}^m$ be:

$$m_i(k)=FB_i(c^h{}_i)(z_i(k)).$$

Let

$$m(k)=\underline{FB}(c^h)(z(k))= (FB_1(c^h{}_1)(z_1(k)),\ldots,FB_n(c^h{}_n)(z_n(k))).$$

Then, using the above notations we have

$$z(k+1)=\phi(z(k),\underline{FB}(c^h)(z(k)),\underline{K}'(z(k),\underline{FB}(c^h)(z(k))).$$

The above relation specifies the dynamical representation Π of the over all process under a feedback regulation and is the target of the stability argument and of the optimal management.

5.4 Optimization Activity of Operational Level

Since the optimization activity of the operational level is performed at the coordination planning stage, it will be discussed in relation with the organizational level activity. This section, therefore, introduces its basic framework briefly.

As discussed in Chapter 2, each decision maker D_i of the operational level has an image or a model P_i^m of the target process and his own goal G_i. The goal controls the behavior of the operational level. Specifically, the goal determines the set point or the equilibrium point c^* of the regulation.

The optimization activity of the operational level has two difficulties. The first difficulty is related with feasibility of optimization. Although the operational level is a closed system isolated from the environmental input U_e, each decision maker has uncertainty due to interactions among the sub-processes. Let us consider a trivial example. Suppose the operational level consists of two units. Let the process P be given by

$$\begin{bmatrix} y_1 \\ y_2 \end{bmatrix} = \begin{vmatrix} 1 & 0 \\ 0 & 3 \end{vmatrix} \begin{bmatrix} m_1 \\ m_2 \end{bmatrix} + \begin{vmatrix} 0 & -2 \\ -3 & 0 \end{vmatrix} \begin{bmatrix} y_1 \\ y_2 \end{bmatrix}$$

where every variable is a real. Then, P is closed because it does not have any external input. If P is decomposed into two sub-processes, P_1 and P_2, they are

$$y_1 = m_1 - 2y_2,$$

$$y_2 = 3m_2 - 3y_1.$$

Suppose

$$G_1(y_1) = y_1^2,$$

$$G_2(y_2) = y_2^2.$$

Then, D_1 cannot carry out optimization with respect to his problem unless he knows the value of the interaction input y_2.

It is a responsibility of the upper level management to make the decision problem determinate by providing appropriate information about the uncertainty.

The second difficulty is related with meaning of optimization. In general, there is no guarantee that an optimally calculated behavior of the operational level produces a desirable result in the view of the organizational goal. Then, it may be meaningless for the operational units to pursue optimizing decision. But if the operational level is well coordinated and the organization is valid in the sense of Chapter 1, optimization by the operational level leads to optimization of the organization goal. Significance of the optimization activity of the operational level must be understood in this context.

Let

$$G_i:M_i \times Y_i \rightarrow Re$$

be the goal of the i-th operational level unit. This is a general form of the goal representation where D_i evaluates its outcome y_i in relation to its effort m_i.

Let the model of the i-th sub process be

$$P_i^m:M_i^m \rightarrow Y_i\,.$$

M_i^m may or may not be equal to M_i depending on the coordination policy of the organizational level management which takes care of the uncertainty caused by the interaction. Then, the decision model P_i^m does not include an uncertainty component.

Let

$$H_i^m:M_i^m \rightarrow M_i\,.$$

H_i^m maps a decision of the model to a real manipulating variable of P_i. H_i^m will be called interpretation function.

Let a reduced goal $g_i^m:M^m \rightarrow Re$
be

$$g_i^m(m_i^m)=G_i(m_i,P_i^m(m_i^m)).$$

Let m_i^{m*} be an optimal solution, i.e.,

$$(\forall\, m_i^m \in M_i^m)(g_i^m(m_i^m) \leqq g_i^m(m_i^{m*}))\,.$$

Then,

$$y_i^* = P_i^m(m_i^{m*})$$

is an optimally calculated desirable output.

The decision problem of the operational level unit is in general a dynamic optimization. Since an organization is assumed to be a discrete system, however, there is a way to make its dynamic optimization problem be reduced to a static optimization which is much simpler than a dynamic optimization.

Suppose the optimization operation is to be done on the time scale $T_{coord}=[0,..,n]$ where n is not a large number. Under this assumptions the dynamic optimization problem is represented as follows:

$$z_i^m(1)=\phi_i^m(z_i^m(0),m_i^m(0))$$

$$z_i^m(2)=\phi_i^m(z_i^m(1),m_i^m(1))$$

$$\vdots$$

$$z_i^m(n)=\phi_i^m(z_i^m(n-1),m_i^m(n-1))$$

$$y_i(0)=\lambda_i^m(z_i^m(0))$$

$$y_i(1)=\lambda_i^m(z_i^m(1))$$

$$\vdots$$

$$y_i(n)=\lambda_i^m(z_i^m(n))\ .$$

Let

$$\underline{z}_i^m=[z_i^m(0),..,z_i^m(n)],$$

$$\underline{m}_i^m=[m_i^m(0),..,m_i^m(n)],$$

$$\underline{y}_i=[y_i(0),..,y_i(n)].$$

Then, the original dynamic optimization problem for the i-th unit becomes the following static optimization problem:

$G_i'(\underline{m},\underline{y}_i) \to \max$ subject to the constraint

$$\underline{z}_i^m = \underline{\phi}_i^m(z_i^m(0), \underline{m}_i^m)$$

$$\underline{y}_i = \underline{\lambda}_i^m(\underline{z}_i^m)\,.$$

G_i' is derived from G_i. $\underline{\phi}_i^m$ and $\underline{\lambda}_i^m$ are given by:

$$z_i^m(n) = \underline{\phi}_i^m(\underline{\phi}_i^m(\ldots), m_i^m(n-1)),$$

$$y_i(n) = \lambda_i^m(z_i^m(n)).$$

Inequality constraints can be easily taken care of in the static formulation.
If P_i^m and G_i' are linear, the above formulation yields a linear programming problem, which can be solved by the powerful algorithm of the linear programming.

CHAPTER 6 ORGANIZATIONAL LEVEL MANAGEMENT

COORDINATION ACTIVITY

According to the VSM the organizational level management has three functions: To produce a coordination plan, to supervise its execution and to audit the performance of the operational level over an execution period. This chapter and the following two chapters, focus the investigation on the coordination of the operational level units. This is a main function of the organizational level management. In particular, a theoretical study of the planning activity is reported. The ultimate goal of the organizational level management is to secure implementation of a global goal, which is given by the institutional level, by coordination.

As discussed in Chapter 5, the management function of the operational level is performed by regulation and optimization. The organizational level manages the optimization activity of the operational level, by a negotiation process of coordination planning. The organizational level manages the operational level by directing the optimization activity. The regulation function is controlled by the set point specification of the optimization function.

This chapter will proceed in the following way.

First, the concept of an organizational level management structure is introduced as a sub-model of the overall organization level model. The scope of the model covers the operational level as well as the organizational level.

Second, a goal seeking model of the organizational level is introduced so that its coordination function can be formally represented. The task model (goal seeking activity model) of the operational level used in the organizational level management structure specifies a management environment for the organizational level. The coordination of the organizational level must be designed to be compatible with the environmental condition.

Third, a coordination skeleton, which represents the relation between the two goal seeking activities on the operational level and on the organizational level, is formalized.

Fourth, design criteria are formalized for the coordination skeleton.

Finally, a coordination principle concept and a design criterion are introduced. Coordination principle is a key concept on which a coordination strategy of the organizational level is designed. This chapter introduces that concept in a concrete form. Theoretical results are then presented in Chapter 7 and Chapter 8. Implementation of the coordination skeleton that is compatible with the design criteria of this chapter is a core of the subsequent chapters.

6.1 Total Scheme of Organizational Level Management

Fig. 6.1 shows a model for a coordination plan making activity.

It consists of two levels; the operational and organizational. The operational level consists of n goal seeking systems. The processes of these systems are denoted by $P_1,..,P_n$. They interact with each other through the interaction block denoted by K. The totality of the sub-processes including the interaction block will be called the overall process.

Decision makers, which are denoted by $D_1,..,D_n$, respectively, are associated with n processes $P_1,..,P_n$. The former will be called the operational decision makers. The combination of D_i and P_i will be called the i-th operational unit.

The sub-process P_i is an input/output system whose inputs are the manipulating variable m_i specified by D_i and the interaction input u_i comes from the interaction block. The output of P_i is denoted by y_i.

D_i is characterized by its goal G_i and a process model P_i^m. As discussed below, (P_i^m,G_i) is parameterized by a coordination variable specified by the organizational level. For each instance of the coordination variable, (P_i^m,G_i) becomes a deterministic optimization model. The decision principle of the operational level units is specified as optimization.

The organizational level is represented by one decision maker D_0 which will be called *organizational decision maker*. The main task of D_0 is coordination of the operational units.

The manipulating variable of D_0 is a coordination variable. In organization theory coordination is usually assumed to be performed by allocation of resources. In our model this assumption will be made morestructured. The coordination variable will represent either the cost of a management resources or the resources themselves. The former case indicates an indirect intervention management while the latter indicates a direct intervention by the organizational level.

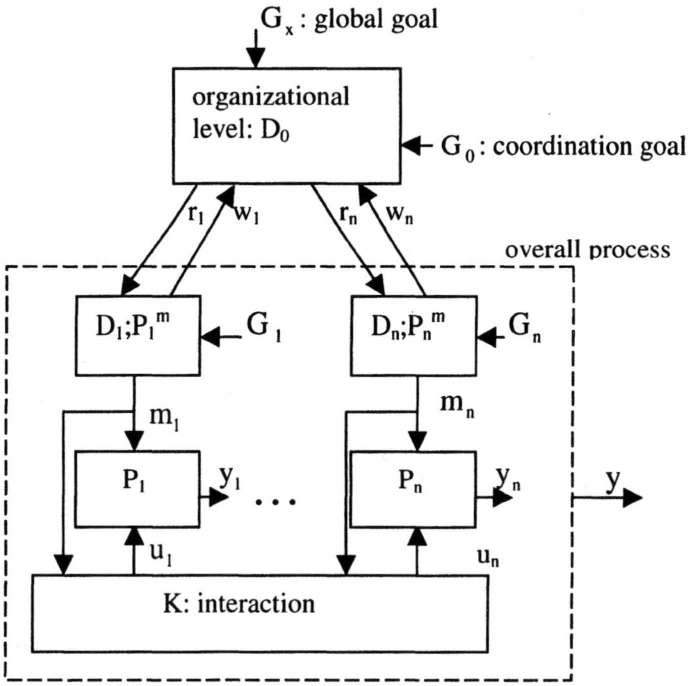

Figure 6.1 Organizational structure

The coordination variable impacting D_i is denoted by r_i. D_0 has two goals, G_0 and G_x. G_x is the global goal, which the institutional level generates for the organizational to pursue as a responsibility of the latter within the organization taking into account the environmental conditions. This will be discussed in Chapter 10 in detail. On the other hand, G_0, which will be called coordination goal is the organizational level used as a criterion to evaluate the state of the coordination state or harmonization of the operational units. G_0 is clearly distinct from the global goal G_x.

Although the overall process is closed, since an interaction input u_i is determined by other operational units, D_i cannot proceed in an optimum fashion based on P_i without information of the activities of other decision determined since P_i is open, i.e., impacted by the uncertainty u_i by other decision makers. As described earlier, the problem of D_i is not well-structured for optimization since u_i is an unknown input to P_i. The coordination r_i provides required information about u_i in order to make the optimization problem of D_i well formed. The goal seeking model of D_i, hence, must include r_i as a parameter. Since the process model of the goal seeking problem of D_i may include r_i, the model, P_i^m can be different from P_i.

Since a coordination planning is a negotiation process, it is an iterative process. After D_i has made an optimum decision based on model specified by a given coordination value r_i, D_i sends an information about its decision to D_0. The information is denoted by w_i in Fig. 6.1, and will be called reporting information. D_0 evaluates the current coordination state using G_0 and the reported information $(w_1,..,w_n)$. If the current state is not satisfactory, D_0 will adjust the current coordination value based on the reporting information. A newly generated coordination values are then sent back to the operational units. The operational units modify its previous decisions using new coordination values. The result is reported to D_0 again. This process is repeated until a satisfactory value of G_0 is achieved.

The interaction block K is an input output system which transforms the input $((m_1,..,m_n),(y_1,..,y_n))$ into $(u_1,..,u_n)$.

This is the total scheme of coordination planning used in this book.

In order to get a better insight into processes and relationships considered in this chapter let us consider the following decision problem where every variable is scalar.

Suppose there are the following three subprocesses:

$P_1{:}y_1=m_1$

$P_2{:}y_2=m_2+u_2;$

$u_2=y_1$

$P_3{:}y_3=m_3+u_3;$

$u_3=y_2$

Fig. 6.2 shows the interaction scheme where u_2 and u_3 are the interaction inputs.

Suppose the global goal is given by:

$$G_x(y)=-(y_1-1)^2-(y_2-2)^2-(y_3-3)^2$$

where $y=(y_1,y_2,y_3)$.

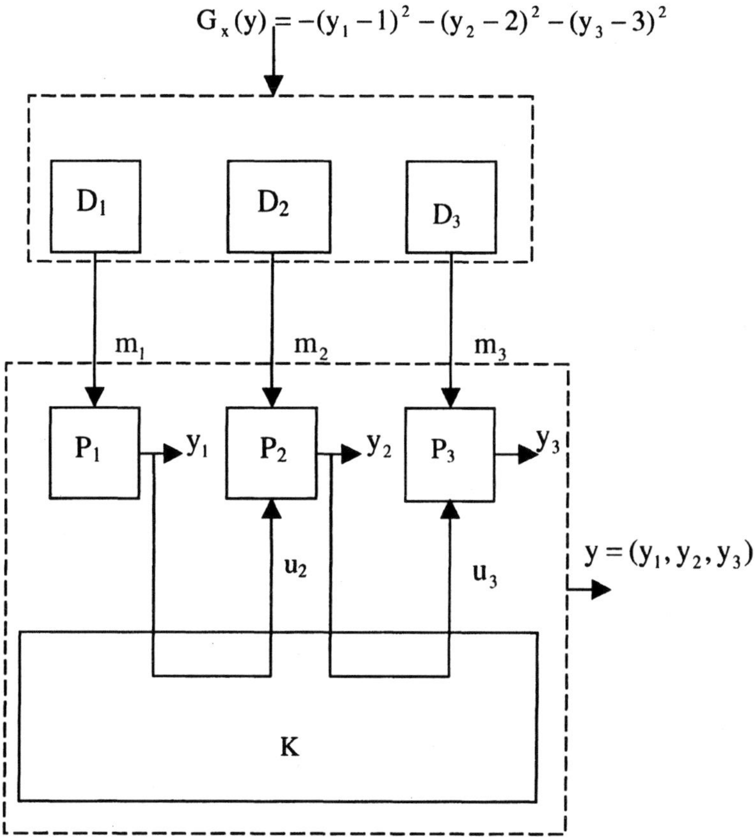

Figure 6.2 Interaction scheme

P=(P_1,P_2,P_3) specifies the overall process and (P,G_x) provides a deterministic optimization problem.

Let us apply the Lagrangean multiplier method to the current optimization problem. Let (γ_1,γ_2) be a multiplier for the equality constraint of the interaction. The Lagrangean formulation is then given as:

$$\begin{aligned}\phi(y,u,\gamma) &= G_x(y)+\gamma_1(y_1-u_2)+\gamma_2(y_2-u_3)\\ &= -(y_1-1)^2+\gamma_1 y_1\\ &\quad -(y_2-2)^2-\gamma_1 u_2+\gamma_2 y_2\\ &\quad -(y_3-3)^2-\gamma_2 u_3\end{aligned}$$

where u=(u_1,u_2,u_3) and γ=(γ_1,γ_2). u_1 is a dummy variable.

Let

$r_1 = \gamma_1$,

$r_2 = (\gamma_1, \gamma_2)$,

$r_3 = \gamma_2$.

Let us define three independent optimization problems as follows:

$D_1(r_1)$:

$P_1^m(r_1)$: $y_1 = m_1$,

$G_1(r_1)(m_1, u_1, y_1) = -(y_1 - 1)^2 + \gamma_1 y_1$.

$D_2(r_2)$:

$P_2^m(r_2)$: $y_2 = m_2 + u_2$,

$G_2(r_2)(m_2, u_2, y_2) = -(y_2 - 2)^2 - \gamma_1 u_2 + \gamma_2 y_2$.

$D_3(r_3)$:

$P_3^m(r_3)$: $y_3 = m_3$,

$G_3(r_3)(m_3, u_3, y_3) = -(y_3 - 3)^2 - \gamma_2 u_3$.

Let the decision variable of $D_i(r_i)$ be:

$m_i^m = (m_i, u_i)$.

Then, the original decision problem can be decomposed into three independent optimization problems for each $r=(r_1,r_2,r_3)$.

Let us derive a management model of the organizational level. Let

$$r=(r_1,r_2,r_3)$$

be a coordination value. γ represents prices of products exchanged between the first level decision units, D_i.

Let $D_1(r_1)$,$D_2(r_2)$ and $D_3(r_3)$ be goal seeking models of the operational level. The i-th operational level decision maker is required to maximize $G_i(r_i)$ with respect to $m_i^m=(m_i,u_i)$ subject to $P_i^m(r_i)$. Notice that in the current model the decision variable m_i^m is different from the real manipulating variable m_i.

Let

$$m^{m^*}(r)=[m_1^{m^*}(r_1),\ m_2^{m^*}(r_2),\ m_3^{m^*}(r_3)]$$

be an optimal solution for a given coordination value r and let it be the reporting information from the operational level to the organizational level decision maker D_0 where $m_i^{m^*}(r_i) = [m_1^*(r_i),\ u_i^*(r_i)]$.

Let the coordination goal G_0 of D_0 be given by a coordination principle called IBP (interaction balance principle) which will be discussed in Chapter 7 in detail; i.e., G_0 is considered to be achieved if and only if the following holds:

$$u_2^*(r_2) = y_1^*(r_1)$$

and

$$u_3^*(r_3) = y_2^*(r_2)$$

where $y_i^*(r_i) = P_i(m_i^{m^*}(r_i))$.

The direct objective of management of D_0 is to find a coordination value r which satisfies the above goal. As mentioned above, search for the desirable coordination value is carried out in an iteration (negotiations between the organizational level and the operational level) process.

In the above organizational level management model, the validity condition can be shown to be satisfied, that is,

$$\text{attainment of } G_0 \rightarrow \text{attainment of } G_x$$

holds.

As a matter of fact, let

$$\phi'(m,u,r) = G_1(r_1)(m_1,u_1,P_1{}^m(r_1)(m_1,u_1))$$

$$+ G_2(r_2)(m_2,u_2,P_2{}^m(r_2)(m_2,u_2))$$

$$+ G_3(r_3)(m_3,u_3,P_3{}^m(r_3)(m_3,u_3)).$$

Then, ϕ' is concave and convex with respect to (m,u) and r , respectively. Then mini-max theorem of Appendix II implies that

$$\min_r \max_{(m,u)}\phi'(m,u,r)=\max_{(m,u)}\min_r\phi'(m,u,r).$$

If

$$\min_r\max_{(m,u)}\phi'(m,u,r)=\phi'(m^*,u^*,r^*),$$

the following holds:

$$u_2^* = P_1^m(r_1^*)(m_1^*,u_1^*),$$

$$u_3^* = P_2^m(r_2^*)(m_2^*,u_2^*)$$

and

$$\max_m G_x(P(m))=G_x(P(m^*)),$$

that is, m^* is an optimum solution of the original problem.

The above result of the min-max theorem implies that a desirable r^* is characterized by

$$\min_r \max_{(m,u)}\phi'(m,u,r)= \max_{(m,u)}\phi'(m^*(r),u^*(r),r).$$

That is, r^* is characterized as a value to minimize $\phi'(m^*(r),u^*(r),r)$, where $m^*(r)$ and $u^*(r)$ are reports from the operational level.

6.2 Formulation of Organizational Level Management

The overall process and the global goal constitute environmental conditions for design of the organizational level management scheme. This chapter will introduce the environmental conditions in a simplified form where an initial state of the overall process is assumed fixed.

Definition 6.1 Environment condition of organizational level management

(i) Overall process:

Let the manipulating variable of the overall process be:

$$M = M_1 \times M_2 \times \cdots \times M_n.$$

Let the output variable of the overall process be:

$$Y = Y_1 \times Y_2 \times \cdots \times Y_n.$$

Let the interaction variable be:

$$U = U_1 \times U_2 \times \cdots \times U_n.$$

Let the i-th subprocess be:

$$P_i : M_i \times U_i \rightarrow Y_i.$$

Let the i-th interaction be:

$$K_i : M \times Y \rightarrow U_i.$$

Let the overall process be:

$$P : M \rightarrow Y.$$

Let

$$K(m)=(K_1(m,P(m)),..,K_n(m,P(m)))$$

and

$$K'_i(m)=K_i(m,P(m)).$$

(ii) Global goal:
Let the global goal G_x be:

$$G_x : Y \rightarrow Re.$$

Let g_x:M $\rightarrow$ Re be:

$$g_x(m)=G_x(P(m)).$$

The *environmental condition* (P,G_x) specifies a deterministic optimization problem. Then let the set of optimum solutions of (P,G_x) be denoted by $Global_{opt}$, i.e.,

$$Global_{opt}=\{m^*|(\forall m)(m \in M \rightarrow g_x(m) \leqq g_x(m^*))\}.$$

Next, we will formalize a coordination model. As mentioned above, a coordination variable will be specified to represent a management resource or a cost of a management resource. The next definition will, however, introduce it in an abstract general form.

Definition 6.2 Coordination model for organizational level management

(i) Operational level goal seeking model
Let the coordination variables for the i-th operational level unit be:

R_i = set of the coordination values for D_i.

Let

$$R = R_1 \times \cdots \times R_n.$$

Let the process model of D_i be:

$$P_i^m(r_i):M_i^m(r_i) \rightarrow Y_i$$

where $M_i^m(r_i)$ is the set of values of the manipulating variable of the goal seeking model of the i-th operational unit which, in general, depends on the coordination value r_i. Notice that $P_i^m(r_i)$ does not include an interaction variable. It is assumed that an information about the interaction is included in r_i.

Let the goal of D_i be:

$$G_i(r_i):M_i^m(r_i) \times Y_i \rightarrow Re$$

where $G_i(r_i)$ depends on the value r_i.

$(P_i^m(r_i),G_i(r_i))$ constitutes a deterministic optimization problem for each r_i. That is, when r_i is given, D_i makes an optimal decision using $(P_i^m(r_i),G_i(r_i))$.

Let $m_i^{m^*}(r_i)$ denote an optimum solution of $(P_i^m(r_i),G_i(r_i))$.

Let

$$M^m(r)=M^m{}_1(r_1)\times\cdots\times M^m{}_n(r_n)$$

and

$$m^{m^*}(r)=(m_1^{m^*}(r_1),.., m_n^{m^*}(r_n)).$$

Let, in general,

$$W_i = \text{set of the reporting values from } D_i.$$

and

$$W = W_1 \times\cdots\times W_n.$$

$m^{m^*}(r)$ will be used for the reporting in this book.

(ii) Interpretation of $M_i^m(r_i)$

Since $M_i^m(r_i)$ is, in general, different from the real manipulating variable M_i, there should be an interpretation of $M_i^m(r_i)$ related to the real manipulation variable in order to implement the result of the decision activity of D_i.

Let

$$M_i^h = \text{superset of } M_i,$$

$$M^h = M_1^h \times M_2^h \times\cdots\times M_n^h,$$

$$H_i^m: \bigcup M_i^m(r_i) \rightarrow M^h{}_i$$

or

$$H^m: \bigcup M^m(r) \to M^h.$$

A coordination variable is essentially a representation of interaction and as the example of Sec. 6.1 shows, there could be cases where $M_i^m(r_i)$ includes U_i. Then, let H^u be:

$$H_i^u : R \times \bigcup M_i^m(r_i) \to U_i$$

or

$$H^u : R \times \bigcup M^m(r) \to U.$$

M is in general given as a subset of M^h satisfying a constraint given by the management resources, which is not explicitly introduced in the current formulation. $H^m \times H^u$ will be called interpretation of $M^m(r)$.

(iii) Organizational level model

Since the reporting information w_i is given by $m_i^{m*}(r_i)$, i.e.,

$$w_i = m_i^{m*}(r_i),$$

the process model of D_0 for the negotiation process is given as $P_0 : R \to \bigcup M^m(r)$ such that

$$P_0(r) = m^{m^*}(r).$$

The process of D_0 is an input-output system whose input is the coordination value to the operational level and whose output is the reporting information from it.

Let the organizational level goal G_0 of D_0 be:

$$G_0 : R \times \bigcup M^m(r) \to \{\text{True}, \text{False}\}.$$

G_0 is represented as a predicate which indicates whether or not a selected coordination realizes a harmonization of the operational units. Concrete forms of G_0 will be given after coordination principles are specified in Chapter 7 and Chapter 8.

For the sake of notational convenience we introduce g_0 as follows:

$$g_0 : R \to \{\text{True}, \text{False}\}$$

such that

$$g_0(r)=G_0(r,P_0(r)).$$

g_0 explicitly indicates whether or not a coordination r is satisfactory.

6.3 Example of Formulation

Let us examine the example of Section 6.1 in the present formulation. The manipulating variable is:

$$M = M_1 \times M_2 \times M_3.$$

The output variable is:

$$Y = Y_1 \times Y_2 \times Y_3.$$

The interaction variable is:

$$U = U_1 \times U_2 \times U_3.$$

The subprocesses are:

$$P_1(m_1,u_1) = m_1.$$

$$P_2(m_2,u_2) = m_2+u_2.$$

$$P_3(m_3,u_3) = m_3+u_3.$$

The interaction inputs are:

$$K_1(m,y) = \text{empty}.$$

$$K_2(m,y) = y_1.$$

$$K_3(m,y) = y_2.$$

The overall process is given by the solution of the following simultaneous equations:

$y_1 = m_1.$

$y_2 = m_2 + y_1.$

$y_3 = m_3 + y_2.$

Or

$y_1 = m_1.$

$y_2 = m_2 + m_1.$

$y_3 = m_3 + m_2 + m_1.$

K_i' is given by:

$K_1'(m) = \text{empty}.$

$u_2 = K_2'(m) = m_1.$

$u_3 = K_3'(m) = m_2 + m_1.$

The global goal is:

$G_x(y) = -(y_1-1)^2 - (y_2-2)^2 - (y_3-3)^2$

or

$g_x(m) = -(m_1-1)^2 - (m_2+m_1-2)^2 - (m_3+m_2+m_1-3)^2.$

The goal seeking model of D_i is:

$m_i^m = (m_i, u_i)$

and

$$P_i^m(r_i)(m_i^m) = P_i(m_i,u_i).$$

The goal of D_i is:

$$G_1(r_1)(m_1,u_1,y_1) = -(y_1-1)^2 + \gamma_1 y_1,$$

$$G_2(r_2)(m_2,u_2,y_2) = -(y_2-2)^2 - \gamma_1 u_2 + \gamma_2 y_2,$$

$$G_3(r_3)(m_3,u_3,y_3) = -(y_3-3)^2 - \gamma_2 u_3.$$

The interpretation is:

$$H^m((m_1,u_1),(m_2,u_2),(m_3,u_3)) = (m_1,m_2,m_3),$$

$$H^u(r,((m_1,u_1),(m_2,u_2),(m_3,u_3))) = (u_1,u_2,u_3).$$

The organizational goal is:

$$g_0(r)=\text{True} \leftrightarrow (\forall\, i)(K_i'(m^*(r))=u_i^*(r_i)).$$

As mentioned in Sec. 6.1, $g_0(r)$ = True, implies that $m^*(r) = m^*$ which is an optimum solution of the original problem. This is not a trivial fact.

6.4 Coordination Principle and Design of Organizational Level Management

a. Concept of Coordination Principle

The function of the organizational level is to make a plan to coordinate activities of the operational level (using coordination variable) and implement the plan hopefully to realize the global goal.

As the formulation of Sec. 6.2 shows, $(r_i,P^m{}_i(r_i),G_i(r_i),g_0)$ is a design parameter of the coordination structure. In this section we will introduce a coordination principle concept which determines a coordination strategy providing a basis for the selection of coordination parameters.

Coordination is necessary even if there are not external inputs because there are interactions among the operational goal seeking units. If there is no interaction and hence if the decision making models are independent from each other, the operational units can make optimum decisions as independent goal seeking systems and the coordination by the organizational level is not needed. Of course, whether or not the global goal is satisfied in such a case by the optimum decision is another story.

Conversely, it can be immediately deduced from this fact that the coordination scheme must depend on how interactions are treated. It will be shown that coordination principles introduced subsequently are named after the way the interactions are treated.

So far several coordination principles have been proposed [2]. This book will introduce two principles, IBP (interaction balance principle) and IPP (interaction prediction principle).

In the IBP the prices of resources are used as coordination variables and the operational level decision makers are to adopt their optimal strategies in order to properly handle the interactions, that is, to maximize their objective with respect to the interactions as well as to the local decision variables. They assume that the values of interactions can be consistent with their own optimal requirements and make the decisions accordingly. The IBP requires then that the selection of the coordination value should be made so as to realize the condition that the interaction assumed by the operational decision makers ("desired" interactions) is equal to the real interaction which is generated when the optimally calculated decision is implemented.

Of course this balanced situation can be achieved only after a negotiation (an iterative information exchange) process is executed between the operational and the organizational levels. Furthermore, it is important to notice that realization of the balance between the assumed and the real interactions does not necessarily imply that the global goal is achieved or the validity condition is satisfied. Therefore, two fundamental questions have to be solved about the coordination process: Convergence problem of the iteration process; and Realization problem for the validity condition. Since this book is concerned with conceptual and structural aspects of an organization, the convergence problem will not be discussed [8].

The other coordination principle, the IPP, is completely different from the IBP in the sense that the operational level decision makers do not address the question of how to deal with the unknown interaction. Rather, the organizational level decision maker predicts interaction values among the operational units using the reporting information from the operational level. The predicted interaction value is used as a coordination information in the IPP. Since the interaction usually represents a resource or a product, a coordination variable in the IPP is a resource or a product. The operational units carry out optimization using the information.

Then, the IPP requires that the coordination process ultimately satisfies the condition that the predicted interaction (by the organizational decision maker) is equal to the real interaction which should be generated when the decision value from the prediction is implemented. It is also true as in the case of the IBP that this condition can be achieved only after an iterative information exchange process and that the validity condition is not necessarily satisfied even when the required coordination situation is realized.

The IPP is conceptually simpler than the IBP. In general it is not so obvious how to define an effective coordination variable in the IBP case. In reality the coordination variable takes a various forms. Since coordination is necessary because of existence of interactions among the operational units, a prediction of the interaction itself is clearly a first choice for the coordination variable. The IPP seems to be more intuitive. When the IPP is used, the required role of the organizational level is simple, that is, to predict a precise value of the interaction. This simplicity may be a main reason why many examples of the IPP are observed in practice. The simplicity, however, does not mean an easy realization of the validity condition as Chapter 8 shows.

b. Design Criteria for Coordination Scheme

As the above formulation indicates, $(r_i, M_i^m(r_i), P_i^m(r_i), G_i(r_i), g_0)$ must be specified in order to implement an organizational level management scheme. The design criteria are given by the following three conditions: Consistency condition, coordinability requirement and validity condition.

Definition 6.3 Design criteria for coordination scheme

Consistency condition of $m^m \in UM^m(r)$

m^m is said to satisfy behavior consistency with respect to r if and only if

$$H^u(r, m^m) = K(H^m(m^m)).$$

Coordinability requirement

The following is called behavior coordinability condition:

$$(\exists\, r)(H^u(r, m^{m^*}(r)) = K(H^m(m^{m^*}(r)))).$$

Validity condition of coordination scheme

The following is called validity condition:

$$g_0(r) = \text{True} \rightarrow H^m(m^{m^*}(r)) \in \text{Global}_{opt}.$$

The behavior consistency means that the interpreted interaction $H^u(r,m^m)$ of m^m is equal to the real interaction generated by applying $H^m(m^m)$ to the real process P. The coordinability requirement implies that there should be a coordination value for which the corresponding consistency holds. Realizability of coordination is required.

A precise definition of the validity condition is given as above. The validity condition is a resolution condition of the conflict between the organizational goal and the operational level goals. Furthermore, it can be interpreted as a condition which links the micro level behaviour of an organization with its macro level behavior. The coordination evaluated by g_0 is concerned with the optimal primary activity or the operational level behavior, that is, with the micro level behavior. On the other hand, the global goal is concerned with the overall performance of the organization, that is, with a macro level behavior.

c. Design Problem of Organizational Level Management Structure

The final target of investigation of an organizational level management structure is to find conditions which makes the organizational structure satisfy the validity condition.

Two cases will be investigated in the subsequent chapters.

The first case is that sub-processes and interactions among them are assumed given (fixed) and the main problem is to design a desirable $(r_i,G_i(r_i))_i$. That is, the task of an operator is well-defined and he can use the real process model in his decision making model. The design problem for the organization level is to find a coordination variable and appropriate goals of the operational level decision making model. We call this case coordination by (operational) goal assignment and will discuss this case in Chapter 7 adopting the IBP.

The second case is that each operational decision maker has its own fixed goal $G_i:Y_i \rightarrow Re$. The goal is used as the goal of his decision making model. In this case the problem is to design $(r_i,P^m{}_i(r_i))_i$. That is, each operator has his own value system and the problem is to assign an appropriate job model to him. We call this case coordination by (operational) task assignment. This case will be discussed in Chapter 8 adopting the IPP.

CHAPTER 7 GOAL ASSIGNMENT COORDINATION

INTERACTION BALANCE PRINCIPLE

In this chapter we will discuss the following design problem for an organizational level management structure: Given the global goal, the overall process and the sub-processes of the operational level, find the coordination variable and the operational level goals so that the generated organization structure satisfies the validity condition for the interaction balance coordination principle IBP.

As shown in Chapter 6, the design parameters of the organization structure are $(r_i, P_i^m(r_i), G_i(r_i), g_0)$. When the IBP is used and when the sub-processes are given, $P_i^m(r_i)$ and g_0 are given by the very definition of the IBP. Then the real problem of this chapter is to find desirable r_i and $G_i(r_i)$.

In other words, the design problem of this chapter—as specified by the IBP concept—is: After accepting the operational procedure (process) of the operational level as given, generate the appropriate operational level goals so as to ensure the IBP coordinability conditions.

Sec. 7.1 will present the formulation of the organizational structure for the IBP under the current assumption.

Sec. 7.2 will introduce an intra-level function for the IBP parameterized by the coordination variable.

Sec. 7.3 will present theorems which will characterize a valid structure for the IBP and coordinability for the IBP in reference to the intra-level function of Sec. 7.2. The validity is given in terms of goal compatibility, while the coordinability condition is given in terms of a min-max theorem,. The results show that a successful coordination by the IBP is implied by a monotonic relation between the global goal and the operational level goals. This certainly agrees with intuition.

Sec. 7.4 will discuss a design procedure for the search for the coordination variable and its corresponding operational level goals.

7.1 Organizational Level Management Structure on IBP (Interaction Balance Principle)

Let

$$G_x:Y \rightarrow Re$$

be the global goal.

Let

$$P_i:M_i \times U_i \rightarrow Y_i$$

be the i-th sub process.

Let

$$P:M \rightarrow Y$$

be the overall process.

Let

$$K_i:M \times Y \rightarrow U_i$$

be the i-th interaction input.

Let

$$u_i=K_i'(m)=K_i(m,P(m))$$

and

$$K(m)=(K_1'(m),..,K_n'(m)).$$

Suppose the above G_x, P_i, P and K_i' are fixed. Then R_i, the set of the i-th coordination values and $G_i(r_i):M_i \times U_i \times Y_i \rightarrow Re$, the i-th operational level goal are the design parameters for the organizational level management structure.

A formulation of the IBP organizational level structure is given below.

Definition 7.1 IBP organization structure

(i) Operational level model for IBP

Let the decision variable of the i-th operational level unit be

$$M_i^m = M_i \times U_i.$$

Let the process model of the i-th operational level unit be

$$P_i^m : M_i^m \rightarrow Y_i$$

where

$$P_i^m(m_i,u_i)=P_i(m_i,u_i).$$

P_i^m is independent of r_i.

Let the goal of the i-th operational level unit be

$$G_i(r_i) : M_i^m \times Y_i \rightarrow Re.$$

The operational level goal $G_i(r_i)$ is to be designed as a function of r_i so that the i-th operational level unit can be controlled by the organizational level decision maker through the coordination variable r_i.

The decision model of D_i is:

$$\Delta^m{}_i(r_i)=\langle M_i^m,\ P_i^m, G_i(r_i)\rangle.$$

The decision principle of D_i is the max principle with respect to M_i^m. Let an optimum decision $m_i^{m^*}(r_i)$ be

$$\max G_i(r_i)(m_i^m,P_i^m(m_i^m))=G_i(m_i^{m^*}(r_i),P_i^m(m_i^{m^*}(r_i))).$$

Let H^m and H^u be given as in Def. 6.1:

$$H^m((m_1,u_1),..,(m_n,u_n)) = (m_1,..,m_n) \equiv H^m(m,u)$$

and

$$H^u(r,((m_1,u_1),..,(m_n,u_n))) = (u_1,..,u_n) \equiv H^u(r,m,u).$$

Let the reporting information from the i-th operational level unit be

$$w_i(r_i)= m_i^{m^*}(r_i)$$

and

$$w(r)=(m_1^{m^*}(r_1),\ldots, m_n^{m^*}(r_n)).$$

(ii) Organizational level model for IBP

The reduced coordination goal g_0 of the IBP is given as

$$g_0(r)=\text{True} \rightarrow H^u(r,w(r))=K(H^m(w(r)))$$

or

$$g_0(r) = \text{True} \rightarrow u^*(r)=K(m^*(r)).$$

The organizational level management structure on the IBP is essentially characterized by two facts.

1. The interaction input u_i is included as one component of the operational level decision variable.

2. The coordination is done in the way to minimize the gap between $u^*(r)$ and $K(m^*(r))$. The example discussed in Chapter 6 is a typical case of the IBP coordination.

7.2 Intra-level Function for IBP Coordination

Chapter 3 introduced the concepts of intra-level function to characterize a relationship between G_x and $\{G_i\}_i$. This section will extend the concept to the situation when G_i is parameterized by a coordination variable r_i.

Let

$$g_i(r_i)(m_i,u_i) \equiv G_i(r_i)((m_i,u),P_i(m_i,u_i)).$$

Then, a relation $F(r) \subset Re^n \times Re$ can be defined as:

$$(v_1,..,v_n,v) \in F(r) \leftrightarrow (\exists m)(v_i=g_i(r_i)(m_i,K'_i(m)) \text{ and } v=g_x(m))$$

where $g_x(m)=G_x(P(m))$.

Let

$$g_i(r_i)(m) \equiv g_i(r_i)(m_i,K'_i(m))$$

and

$$g(r)(m)=(g_1(r_1)(m),\ldots., g_n(r_n)(m)).$$

Then, the following definition can be introduced.

Definition 7.2 IBP intra-level function

If F(r) can be represented as a function from Re^n to Re, i.e., $(v_1,..,v_n,v) \in F(r)$ and $(v_1,..,v_n,v') \in F(r) \rightarrow v=v'$, then F(r) is called *IBP intra-level function*. F(r) may be a partial function.

If an intra-level function exists, the global goal is expressed as a function of the operational goals as follows:

$$g_x(m)=F(r)(g_1(r_1)(m_1,K'_1(m)),..,g_n(r_n)(m_n,K'_n(m))).$$

Let us consider the example of Sec. 6.4. In that example, $g_i(r_i)$ is given by:

$$g_1(r_1)(m_1,u_1)= -(m_1-1)^2+\gamma_1 m_1,$$

$$g_2(r_2)(m_2,u_2)= -(m_2+u_2-2)^2-\gamma_1 u_2+\gamma_2(m_2+u_2),$$

$$g_3(r_3)(m_3,u_3)= -(m_3+u_3-3)^2-\gamma_2 u_3.$$

Then, since

$$g_1(r_1)(m_1,u_1)+g_2(r_2)(m_2,u_2)+g_3(r_3)(m_3,u_3)$$

$$= -(m_1-1)^2-(m_2+u_2-2)^2+\gamma_1(m_1-u_2)-(m_3+u_3-3)^2+\gamma_2(m_2+u_2-u_3),$$

we have

$$g_x(m)=g_1(r_1)(m_1,K_1'(m))+g_2(r_2)(m_2,K'_2(m))+g_3(r_3)(m_3,K'_3(m)).$$

Consequently,

$$F(r)(v_1,v_2,v_3)=v_1+v_2+v_3.$$

In this case an intra-level function exists. Furthermore it is a monotone function. Improvement of the operational level goal implies that of the global goal. Sec. 7.3 will show that this monotonic property is essential for realization of the validity condition of the IBP.

7.3 Validity Theorem of Operational Goal Assignment Coordination

Using the above formulations this section will show a validity theorem for the IBP. The following result presents a basic condition of a valid organization whose coordination scheme is designed by an operational level goal assignment.

Proposition 7.1 [2,8]
Suppose $(r_i,G_i(r_i))$ *satisfies* Def. 7.1. *Suppose the intra-level function* F(r) *of* the IBP *defined in* Def.7.2 *exists, i.e.,*

$$g_x(m)=G_x(P(m))=F(r)(g_1(r_1)(m_1,K_1'(m)),..,g_n(r_n)(m_n,K_n'(m)))$$

where

$$g_i(r_i)(m_i,u_i)=G_i(r_i)((m_i,u_i),P_i(m_i,u_i))$$

and is a strongly monotonic increasing function, i.e.,

$$(\forall i)(v_i \leqq v'_i) \rightarrow F(r)(v_1,..,v_n) \leqq F(r)(v'_1,..,v'_n)$$

and

$$(\forall i)(v_i \leqq v'_i) \text{ and } (\exists i)(v_i < v'_i) \rightarrow F(r)(v_1,..,v_n) < F(r)(v'_1,..,v'_n).$$

Let

$$g(r)(m,u)=F(r)(g_1(r_1)(m_1,u_1),..,g_n(r_n)(m_n,u_n)).$$

Suppose the following relation holds: For any $(m,u) \in M \times U$

$$g_x(m) \geqq \inf_{r \in R} g(r)(m,u).$$

Then, the organizational level management structure on the IBP of Def. 7.1 satisfies the two conditions:

Behavior coordinability condition, i.e.,

$$(\exists r)(u^*(r) = K(m^*(r)))$$

and validity condition, i.e.,

$g_0(r)=\text{True} \rightarrow m^*(r)$ is an optimum solution of the global goal g_x

if and only if $g(r)(m,u)$ *has a saddle point, i.e.,*

$$\max_{(m,u)} \min_r g(r)(m,u)=\min_r \max_{(m,u)} g(r)(m,u).$$

The proof is given in Appendix.

We will use the following facts for the proof of Prop. 7.1.

Lemma 7.1
Suppose an intra-level function F(r) exists. Then

$$\sup_{(m,u)}\inf_r g(r)(m,u) \geqq \sup_m g_x(m).$$

The proof is given in Appendix.

Lemma 7.2
In general the following relation holds:

$$\inf_r \sup_{(m,u)} g(r)(m,u) \geqq \sup_{(m,u)} \inf_r g(r)(m,u).$$

The proof is given in Appendix.

The essences of Prop. 7.1 is provided by two facts: Existence of a strongly monotonic increasing intra-level function and satisfaction of the following saddle point condition:

$$\max_{(m,u)} \min_r g(r)(m,u)=\min_r \max_{(m,u)} g(r)(m,u).$$

Actually, the saddle point condition plays an essential role for the coordinability argument but the validity condition needs only the strong monotonicity of F(r). We have therefore

Proposition 7.2
Suppose an intra-level function F(r) of Prop. 7.1 exists. If F(r) is a monotonic increasing, the validity condition is satisfied.

The proof is given in Appendix.

Appendix II shows that if g(r)(m,u) is concave with respect to (m,u) for each r and convex with respect to r for each (m,u), the above equality between the min-max and the max-min of g(r)(m,u) holds.

7.4 Operational Level Goal Design for IBP

Let us examine the example introduced in Sec.6.4. The examination will present a procedure to specify the coordination variable and to design the operational level goal of the IBP.

Suppose the global goal is given as:

$$G_x(y) = -(y_1-1)^2 - (y_2-2)^2 - (y_3-3)^2.$$

The process and the sub-processes are:

$$y_1 = P_1(m_1,u_1) = m_1,$$

$$y_2 = P_2(m_2,u_2) = m_2+u_2,$$

and

$$y_3 = P_3(m_3,u_3) = m_3+u_3$$

where the interactions are given by:

$$u_2 = y_1 = K_2(m,y),$$

$$u_3 = y_2 = K_3(m,y)$$

or

$$K_2'(m) = K_2(m,P(m)) = m_1$$

and

$$K_3'(m) = K_3(m,P(m)) = m_1+m_2.$$

The overall process is given by:

$$P(m) = (m_1, m_2+m_1, m_3+m_1+m_2).$$

Then, the decision variables of the operational level in the organizational level management structure are:

$M_1^m = M_1 (= Re)$,

$M_2^m = M_2 \times U_2 (= Re^2)$,

$M_3^m = M_3 \times U_3 (= Re^2)$.

The sub-processes of the operational level decision model are

$P_1^m(m_1, u_1) = m_1 = P(m_1, u_1)$,

$P_2^m(m_2, u_2) = m_2 + u_2 = P(m_2, u_2)$,

$P_3^m(m_3, u_3) = m_3 + u_3 = P(m_3, u_3)$.

Let the coordination variable be defined as follows: Let

$(\gamma_1, \gamma_2) \in Re^2$.

Then, let

$r_1 = \gamma_1$,

$r_2 = (\gamma_1, \gamma_2)$,

$r_3 = \gamma_2$.

Let corresponding goals of the operational level decision model be:

$G_1(r_1)(m_1^m, y_1) = -(y_1 - 1)^2 - \gamma_1 y_1$,

$G_2(r_2)(m_2^m, y_2) = -(y_2 - 2)^2 + \gamma_1 u_2 - \gamma_2 y_2$,

$$G_3(r_3)(m_3{}^m, y_3) = -(y_3-3)^2 + \gamma_2 u_3.$$

Let H^m and H^u be:

$$H^m((m_1,u_1),(m_2,u_2),(m_3,u_3))=(m_1,m_2,m_3),$$

and

$$H^u(r,((m_1,u_1),(m_2,u_2),(m_3,u_3)))=(u_1,u_2,u_3).$$

The goal of the organizational level is:

$$g_0(r)=\text{True} \leftrightarrow (u_2^*(r), u_3^*(r))=(y_1^*(r), y_2^*(r))$$

because

$$\begin{aligned} K(H^m(m^{m^*}(r))) &= (K_2'(m^*(r))), K_3'(m^*(r)))) \\ &= (y_1^*(r), y_2^*(r)). \end{aligned}$$

The above definitions imply

$$\begin{aligned} g_1(r_1)(m_1,u_1) &= -(P_1(m_1,u_1)-1)^2 - \gamma_1 y_1 \\ &= -(m_1-1)^2 - \gamma_1 m_1, \end{aligned}$$

$$\begin{aligned} g_2(r_2)(m_2,u_2) &= -(P_2(m_2,u_2)-2)^2 - \gamma_1 u_2 - \gamma_2 P_2(m_2,u_2) \\ &= -(m_2+u_2-2)^2 - \gamma_1 u_2 - \gamma_2(m_2+u_2), \end{aligned}$$

$$\begin{aligned} g_3(r_3)(m_3,u_3) &= -(P_3(m_3,u_3)-3)^2 - \gamma_2 u_3 \\ &= -(m_3+u_3-3)^2 - \gamma_2 u_3. \end{aligned}$$

Let F(r) be:

$$F(r)(v_1,v_2,v_3)=v_1+v2+v_3 .$$

Then,

$$\begin{aligned}
&F(r)(g_1(r_1)(m_1, K_1'(m)), g_2(r_2)(m_2, K_2'(m)), g_3(r_3)(m_3, K_3'(m))) \\
&= g_1(r_1)(m_1, K_1'(m)) + g_2(r_2)(m_2, K_2'(m)) + g_3(r_3)(m_3, K_3'(m)) \\
&= -(m_1 - 1)^2 - \gamma_1 m_1 - (m_2 + u_2 - 2)^2 + \gamma_1 u_2 - \gamma_2(m_2 + u_2) \\
&\quad - (m_3 + u_3 - 3)^2 + \gamma_2 u_3 \\
&= G_x(P(m)) + \gamma_1(u_2 - m_1) + \gamma_2(u_3 - (m_2 + u_2))
\end{aligned}$$

where $u_2=K_2'(m)$ and $u_3=K_3'(m)$. Since

$$K_2'(m)-m_1=m_1-m_1=0$$

and

$$K_3'(m) - (m_2+K_2'(m)) = m_1+m_2-(m_2+m_1) = 0,$$

$$\begin{aligned}
&F(r)(g_1(r_1)(m_1, K_1'(m)), g_2(r_2)(m_2, K_2'(m)), g_3(r_3)(m_3, K_3'(m))) \\
&= G_x(P(m)) = g_x(m).
\end{aligned}$$

The present F(r) is a strongly monotonic increasing intra-level function. Since

$$g(r)(m,u)=g_x(m)+\gamma_1(u_2-m_1)+\gamma_2(u_3-(m_2+u_2)), \qquad \text{(K)}$$

we have that if $u_2=m_1$ and $u_3=(m_2+u_2)$

$$g_x(m)=g(r)(m,u)$$

and furthermore

$$g_x(m) > \inf_r g(r)(m,u)$$

because g(r)(m,u) is linear with respect to (γ_1, γ_2). Therefore, the condition $g_x(m) \geqq \inf_{r \in R} g(r)(m,u)$ of Prop. 7.1 holds.

Finally, since g(r)(m,u) is concave with respect to (m,u) (because it is of a quadratic form in (m,u)) and convex with respect to r (because it is of a linear form in r), the min-max theorem holds.

Consequently, the organization structure defined above satisfies the coordinability and the validity conditions.

The essence of the design procedure mentioned above is revealed in (K). As mentioned, (K) is a Lagrangean formulation to mitigate the constraint condition of the interaction relation. This can be generalized.

Proposition 7.3

Suppose all variables are defined on finite dimensional real vector spaces.

Let the global goal be:

$$G_x : Y \rightarrow Re.$$

Suppose the sub-processes are given by the following form:

$$y_i = P_i(m_i, u_i)$$

where $u_i = (y_1, \ldots, y_{i-1}, y_{i+1}, \ldots, y_n)$ *or*

$$u_i = \Sigma_{j \neq i} K'_{ij}(y_j)$$

and K'_{ij} *is a matrix of an appropriate dimension.*

Let

$$G_i^0(u_i, y_i) = G_x(y_1', \ldots, y_{i-1}', y_i, y_{i+1}', \ldots, y_n')/n$$

where $u_i = (y_1', \ldots, y_{i-1}', y_{i+1}', \ldots, y_n')$.

Let the goal of D_i *be*

$$g_i(r_i)(m_i, u_i) = G_i(u_i, P_i(m_i, u_i)) + \gamma_i(u_i) - \Sigma_j \gamma_j K'_{ji}(m_i, P_i(m_i, u_i))$$

where

$$r_i = (\gamma_1, \ldots, \gamma_{i-1}, \gamma_{i+1}, \ldots, \gamma_n)$$

and γ_i *is a vector.*

Let $g(r)(m,u)$ *be:*

$$\begin{aligned} g(r)(m,u) &= \sum_i g_i(r_i)(m_i, u_i) \\ &= \sum_i (G_i(u_i, y_i) + \gamma_i(u_i) - \sum_j \gamma_j K'_{ji}(m_i, y_i)) \\ &= \sum_i (G_i(u_i, y_i) + \sum_i \gamma_i(u_i - \sum_{j \neq i} K'_{ij}(m_j, y_j)). \end{aligned}$$

where $y_i = P_i(m_i, u_i)$.

Then, an organizational structure on the IBP *which satisfies the coordinability and the validity conditions can be designed if $g(r)(m,u)$ has a saddle point.*

The proof is given in Appendix.

Notice that the global goal of Prop. 7.3 is not of an additive form.

In general γ_i can be a linear functional on a function space. Since min-max theorem can be extended to a function space, the above presents a fairly general design procedure.

APPENDIX OF CHAPTER 7

Proof of Lemma 7.1

Notice that

$$g(r)(m,K(m)) = F(r)(g_1(r_1)(m_1,K_1'(m)),\ldots,g_n(r_n)(m_n,K_n'(m)))$$
$$= g_x(m) \text{ for any r and m} \qquad \text{(A)}$$

due to the definition of the intra-level function. Therefore, since

$$\sup_u \inf_r g(r)(m,u) \geqq \inf_r g(r)(m,u)$$

for any $(m,u) \in M \times U$, we have

$$\sup_u \inf_r g(r)(m,u) \geqq \inf_r g(r)(m,K(m))$$

for any m. Since

$$\sup_m \sup_u \inf_r g(r)(m,u) \geqq \sup_u \inf_r g(r)(m,u)\,,$$

$$\sup_{(m,u)} \inf_r g(r)(m,u) \geqq \inf_r g(r)(m,K(m))$$

holds for any m. Then we have

$$\sup_{(m,u)} \inf_r g(r)(m,u) \geqq \sup_m \inf_r g(r)(m,K(m))$$
$$= \sup_m \inf_r g_x(m) \quad \text{(due to (A))}$$
$$= \sup_m g_x(m).$$

Q.E.D.

Proof of Lemma 7.2

For any r and (m,u)

$$g(r)(m,u) \geqq \inf_r g(r)(m,u)$$

and

$$\sup_{(m,u)} g(r)(m,u) \geqq g(r)(m,u)$$

hold. Then, we have

$$\sup_{(m,u)} g(r)(m,u) \geqq \inf_r g(r)(m,u)$$

for any r and (m,u). Hence,

$$\sup_{(m,u)} g(r)(m,u) \geqq \sup_{(m,u)} \inf_r g(r)(m,u)$$

for any r. Consequently, we have the result.

Q.E.D.

Proof of Proposition 7.1

Notice that

$$\begin{aligned} g(r)(m,K(m)) &= F(r)(g_1(r_1)(m_1,K_1'(m)),\ldots,g_n(r_n)(m_n,K_n'(m))) \\ &= g_x(m) \text{ for any r and m} \end{aligned} \qquad \text{(B)}$$

due to the assumption of the intra-level function.

We will prove the necessary part first. Suppose the validity condition and the behavior coordinability condition hold. The behavior coordinability condition implies that for some r^*

$$u^*(r^*)=K(m^*(r^*))$$

or

$$g_0(r^*)=\text{True}.$$

Then, the validity condition implies that

$$g_x(m^*(r^*))=\max_m g_x(m) \equiv g_x(m^h) \qquad \text{(C)}$$

Consequently,

$$\begin{aligned} \sup_{(m,u)} g(r^*)(m,u) &= \sup_{(m,u)} F(r^*)(g_1(r_1^*)(m_1,u_1),\ldots,g_n(r_n^*)(m_n,u_n)) \\ &= F(r^*)(\sup_{(m_1,u_1)} g_1(r_1^*)(m_1,u_1),\ldots,\sup_{(m_n,u_n)} g_n(r_n^*)(m_n,u_n)) \end{aligned}$$

(due to the monotonicity of F(r))

$$=F(r^*)(g_1(r_1^*)(m_1^*(r_1^*),u_1^*(r_1^*)),..,g_n(r_n^*)(m_n^*(r_n^*),u_n^*(r_n^*)))$$
$$=g(r^*)(m^*(r^*),u^*(r^*))$$
$$=g(r^*)m^*(r^*),K(m^*(r^*)))$$

(because $u^*(r^*)=K(m^*(r^*)))$

$$=g_x(m^*(r^*))$$

(due to (B))

$$=\max_m g_x(m) \qquad \text{(D)}$$

(due to (C)).

Hence

$$\max_m g_x(m) \leqq \sup_{(m,u)} \inf_r g(r)(m,u)$$

(due to Lemma 7.1)

$$\leqq \inf_r \sup_{(m,u)} g(r)(m,u)$$

(due to Lemma 7.2)

$$\leqq \sup_{(m,u)} g(r^*)(m,u)$$

$$=\max_m g_x(m)$$

(due to (D)). Therefore,

$$\min_r \max_{(m,u)} g(r)(m,u) = \max_{(m,u)} \min_r g(r)(m,u) \qquad \text{(E)}$$

Next we will prove the sufficiency. Suppose (E) holds. Due to the assumption of the proposition we have

$$\sup_m g_x(m) \geqq g_x(m) \geqq \inf_r g(r)(m,u)$$

and hence

$$\sup_m g_x(m) \geqq \sup_{(m,u)} \inf_r g(r)(m,u) \qquad \text{(F)}$$

holds. Then, Lemma 7.1 and (F) imply

$$g_x(m^h) = \max_m g_x(m) \leqq \sup_{(m,u)} \inf_r g(r)(m,u)$$
$$\leqq \sup_m g_x(m).$$

Hence,

$$\max_m g_x(m) = \sup_{(m,u)} \inf_r g(r)(m,u). \tag{G}$$

(E) and (G) imply

$$g_x(m^h) = \max_m g_x(m) = \min_r \max_{(m,u)} g(r)(m,u)$$
$$\equiv g(r^*)(m^*(r^*), u^*(r^*)). \tag{H}$$

Then, for any i

$$g_i(r_i^*)(m_i^*(r_i^*), u_i^*(r_i^*)) = \max_{(m_i,u_i)} g_i(r_i^*)(m_i, u_i)$$
$$\geqq g_i(r_i^*)(m_i^h, K_i'(m^h)).$$

In fact

$$\max_{(m_i,u_i)} g_i(r_i^*)(m_i, u_i) = g_i(r_i^*)(m_i^h, K_i'(m^h)) \tag{I}$$

holds. If

$$g_i(r_i^*)(m_i^h, K_i'(m^h)) < g_i(r_i^*)(m_i^*(r_i^*), u_i^*(r_i^*))$$

were true for some i, we should have

$$g_x(m^h) = g(r^*)(m^h, K(m^h)) < g(r^*)(m^*(r^*), u^*(r^*)) \tag{J}$$

due to the strong monotonic property of F(r), but (J) is incompatible with (H). Since (I) implies that $(m_i^h, K_i'(m^h))$ is an optimum solution with respect to r^* or since $m_i^*(r_i^*) = m_i^h$ and $u_i^*(r_i^*) = K_i'(m^h)$ can hold, the behavior coordinability condition is satisfied.

Next suppose $u^*(r^*) = K(m^*(r^*))$ or $g_0(r^*)$=True. Then,

$$g_x(m^h) = g(r^*)(m^h, K(m^h))$$

(due to (B))

$$\leqq g(r^*)(m^*(r^*),u^*(r^*))$$

(due to the monotonic property of F(r))

$$=g(r^*)(m^*(r^*),K(m^*(r^*)))$$

(due to $u^*(r^*)=K(m^*(r^*))$)

$$=g_x(m^*(r^*))$$

(due to (B)).

Since

$$g_x(m^h) \geqq g_x(m^*(r^*))$$

always holds, we have

$$g_x(m^h) = g_x(m^*(r^*)).$$

Therefore, the validity condition is satisfied.

Q.E.D.

Proof of Proposition 7.2

Notice that Lemma 7.1 or the condition(A) holds when F(r) exists. The proof of the validity condition part of Prop. 7.1, consequently, holds under the present conditions.

Q.E.D.

Proof of Proposition 7.3

The basic idea of the proposition is based on the reformulation of the global goal as follows:

$$\begin{aligned} G_x(y) &= G_x(y)/n+\cdots+ G_x(y)/n \\ &= G_1^0(u_1,y_1)+\cdots+G_n^0(u_n,y_n). \end{aligned}$$

Then, it should be clear that the above formulation has a strongly monotonic intra-level function F(r) such that

$$F(r)(v_1,..v_n)=v_1+\cdots+v_n.$$

Then, we have the desired result from Prop. 7.1.

Q.E.D.

CHAPTER 8 TASK ASSIGNMENT COORDINATION

INTERACTION PREDICTION PRINCIPLE

This chapter is concerned with the following problem: Given the global goal and the operational level goals, find the operational level process or design tasks of the operational level units so that the organizational level management structure based on the IPP satisfies the validity condition.

As mentioned in Chapter 6, the design parameters of the organizational level management structure are r_i, $M_i^m(r_i)$, $P_i^m(r_i)$, $G_i(r_i)$ and g_0. If an operational level goal is given and the IPP is used for coordination, $P_i^m(r_i)$ is the only parameter to be specified for the organizational level design because r_i, $M_i^m(r_i)$ and g_0 are determined automatically by the definition of the IPP when $P_i^m(r_i)$ is given.

This chapter assumes that the compatibility between the organizational goal and the operational level goals is characterized by an organizational culture as presented in Chapter 3. Then, it will be shown that if the culture is a "conducive" one, in particular, if it satisfies a monotonic property, a valid organizational level management structure can be designed for the organization.

Sec. 8.1 will present a formulation of the organizational level management structure under the conditions of this chapter.

As an illustration of the conditions when the IPP leads to a valid management structure, the so-called team decision situation is presented in Sec. 8.2. In that case the operational level goals are design parameters while the overall process is given and hence, every operator is designed to have the same goal or the differences among individual value systems are ignored.

This is an idealized situation since in an organization, indeed in almost any group situation, each operational unit has a goal of its own. The task of the IPP is precisely to harmonize the behavior by coordination. This is a much more difficult situation than considered in the team decision case and is the focus of this chapter.

Sec. 8.3 will present a main theorem for the IPP. The validity condition for the IPP is stated in terms of properties of the organization culture in the main theorem in contrast to the team decision case where the conditions are given in terms of properties of the operational level goals.

8.1 Organizational Level Management Structure in Coordination by IPP

Let

$$G_x: Y \rightarrow Re$$

be the global goal.

Let

$$Y = Y_1 \times \cdots \times Y_n$$

be the output.

Let

$$G_i: Y_i \rightarrow Re$$

be the i-th operational level goals. The operational level is supposed to be concerned only with the outcome of its process, i.e., it has a market driven goal.

Suppose G_x, Y, $\{Y_i\}_i$ and $\{G_i\}_i$ are fixed.

Let

$$\underline{G} = (G_1, .., G_n).$$

In the subsequent arguments

$$Spec = \langle (Y_i)_i, (G_i)_i, G_x \rangle$$

will be referred to as a *specification of an organizational level management structure design* for the IPP.

Under the above notations let us specify the organizational level management structure for the IPP as below.

Definition 8.1 Organizational level management structure on IPP

Let us start from the assumption that M and P:M → Y are already given. Their design will be discussed in is Sec. 8.3. Let

$$M=M_1 \times \cdots \times M_n.$$

Let

$$U_i=M_{-i}=M_1 \times \cdots \times M_{i-1} \times M_{i+1} \times \cdots \times M_n$$

or

$$u_i=(m_1,..,m_{i-1},m_{i+1},..,m_n).$$

Then, the following model is obtained.

(i) Operational level model for IPP
Let the i-th coordination variable be:

$$R_i = U_i.$$

Let the i-th process model be:

$$M_i^m(r_i) = M_i$$

and

$$P_i^m(r_i) : M_i^m(r_i) \rightarrow Y_i$$

or

$$P_i^m(r_i) : M_i \rightarrow Y_i$$

such that

$$P_i^m(r_i)(m_i) = \Pi_i P(m_i,r_i) \equiv P_i(m_i,r_i).$$

where Π_i is the i-th projection operator when

$$r_i=(m_1,..,m_{i-1},m_{i+1},..,m_n) \ (\equiv m_{-i}),$$

$$(m_i,r_i) \equiv (m_1,..,m_i,..,m_n) \ (=m)$$

and

$$P(m_i,r_i) \equiv P(m_1,..,m_i,..,m_n).$$

Although, strictly speaking, m is not equal to (m_i,r_i), the notation $P(m_i,r_i)$ will be used for P(m) to make the relation between P and P_i clear.

Let

$$K_i'(m)=m_{-i}$$

and

$$K(m)=(K_1'(m),..,K_n'(m)).$$

The goal of the i-th operational unit is

$$G_i: Y_i \rightarrow Re.$$

Let

$$g_i(r_i)(m_i) = G_i(P_i(m_i, r_i)) \\ \equiv g_i(m_i, r_i).$$

The decision model of D_i is:

$$\Delta^m(r_i)=<M_i,P_i^m(r_i),G_i>.$$

The decision principle of D_i is the maximization principle with respect to $M_i^m=M_i$.

The interpretation H^m and H^u are

$$H_i^m(m_i)=m_i$$

and

$$H_i^u(r,m_i)=r.$$

The reporting information $w_i(r_i)$ from the i-th operational unit is

$$w_i(r_i)= m_i^*(r_i).$$

(ii) Organizational level model for IPP

Since the coordination principle is the IPP, the coordination g_0 is given as

$$g_0(r) = \text{True} \leftrightarrow r = K(H^m(m^*(r)))$$

where $m^*(r)=(m_1^*(r_1),\ldots,m_n^*(r_n))$ and $m_i^*(r_i)$ is an optimum solution of $\Delta^m(r_i)$.

The organizational structure of the IPP is characterized by the following.

1. The interaction input is used as a coordination variable.
2. A value of the interaction is predicted by the organizational level as a coordination activity so that the behavior consistency condition is realized.

The following facts are of conceptual importance.

Corollary 8.1

Suppose SubSpec=<$\{Y_i\}_i,\{G_i\}_i$> is given. Then, if $P{:}M_1 \times\cdots\times M_n \to Y$ is specified, <SubSpec,P> determines a unique operational level model for the IPP.

Proposition 8.1

The organizational level management structure of Def. 8.1 satisfies the following: For any decision m^m of the operational level

$$H^m(m^m) \in M.$$

The proof is given in Appendix.

8.2 Example of Organizational Level Management Structure on IPP

As an example of the IPP coordination let us consider an application of the team decision theory to an organization where the operational level is considered as a team. This consideration will yield a valid IPP organizational scheme of a specific form.

Suppose the original system is given as in Fig. 8.1.

Let:

$$G_x{:}Y \to Re$$

be the global goal.

In the team decision theory the overall process P is assumed to be known by every decision maker. That is, let

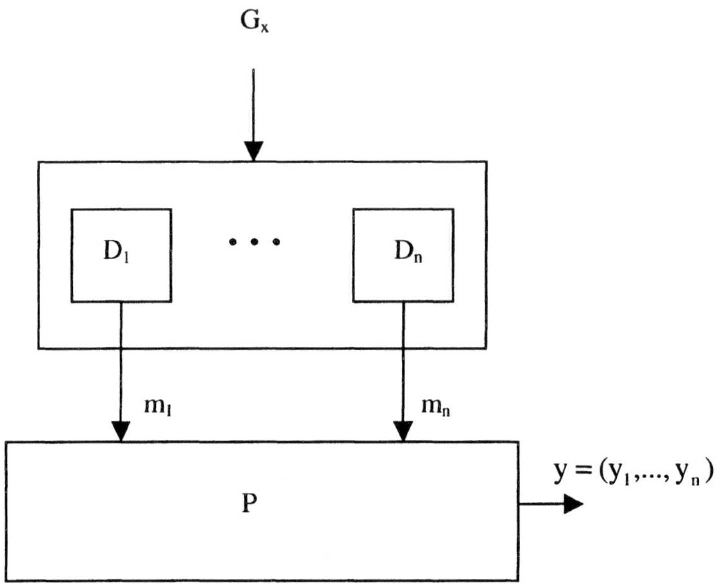

Figure 8.1 Team decision theory

$$P: M \to Y$$

be the overall process.
Let g_x:M → Re be

$$g_x(m)=G_x(P(m)).$$

Let

$$M=M_1 \times \cdots \times M_n$$

and

$$Y=Y_1 \times \cdots \times Y_n.$$

Let us formulate the decision making models of the operational level and of the organizational level following the team decision theory.

Let the decision variable of the decision maker D_i of the operational level be m_i. Let

$$u_i=(m_1,..,m_{i-1},m_{i+1},..,m_n)$$

and

$$r_i=u_i.$$

Let $P_i^m(r_i):M_i \to Y_i$ be such that

$$P_i^m(r_i)(m_i)= \Pi_i P(m_i,r_i).$$

Let

$$G_i(r_i)(m_i,y_i)=g_x(m_1,..,m_{i-1},m_i,m_{i+1},..,m_n)$$

where

$$r_i=(m_1,..,m_{i-1},m_{i+1},..,m_n).$$

Notice that since $G_i(r_i)$ is given as a function independent of y_i, an optimal selection of m_i does not depend on $P_i^m(r_i)$. Hence, $G_i(r_i)(m_i,y_i)$ will be denoted by $G_i(m_i,r_i)$.

Let

$$K_i'(m_1,...,m_n)=(m_1,..,m_{i-1},m_{i+1},..,m_n).$$

The team decision theory does not have an explicit coordinator. However, let D_0 be the organizational decision maker which is supposed to adopt the IPP for the coordination. The coordination goal g_0 of the organizational level is, then,

$$g_0(r)=\text{True} \leftrightarrow (\forall i)(r_i=K_i'(m^*(r))).$$

Fig. 8.2 illustrates the structure of the present coordination scheme.

Suppose the goal g_0 holds for r^*. Then, we have the following condition: For any i and $m_i \in M_i$

$$G_i(m_i, r_i^*)=G_i(m_i, K_i'(m^*(r^*))) \leqq G_i(m_i^*(r_i^*), K_i'(m^*(r^*))).$$

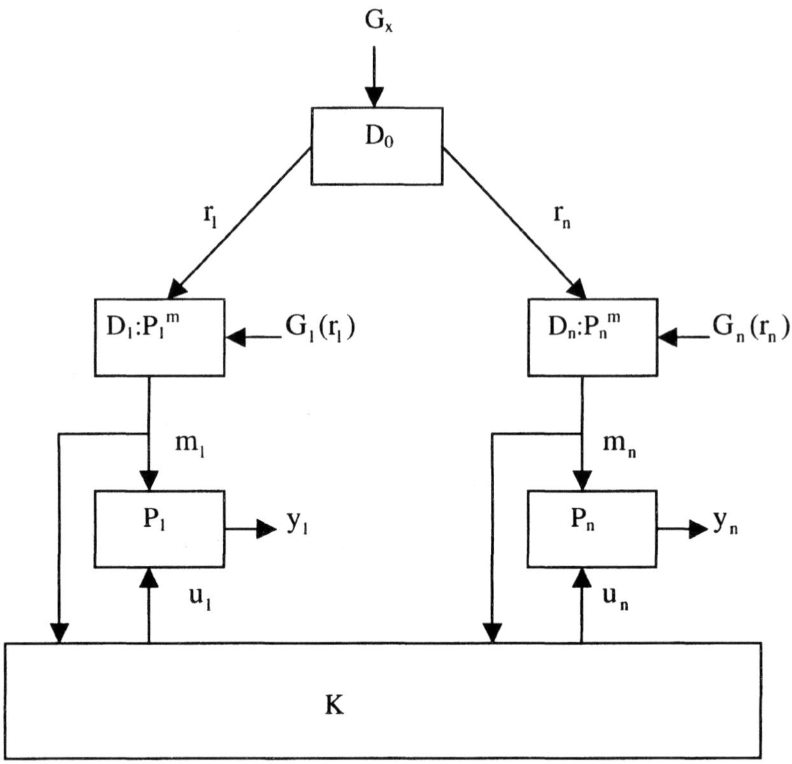

Figure 8.2 Coordination model of team decision theory

$m^*(r^*)$, which is a solution of the IPP coordination, is also a solution of the team decision theory. It is called *person-by-person-satisfactory type* solution (PBPS).

In general, PBPS does not yield an optimal solution of the global goal, i.e., the team decision structure does not satisfy the validity condition of the IPP coordination principle. The following proposition, however, gives a condition under which the validity condition is satisfied for such a case.

Proposition 8.2

Suppose M_i=Re *and* $g_x(m)$ *is a continuously differentiable concave function, i.e.,*

$$g_x(km+(1-k)m) \geqq kg_x(m)+(1-k)g_x(m)$$

where $0 \leqq k \leqq 1$. *Then a PBPS solution* m^* *is a global solution.*

The proof is given in Appendix.

8.3 Validity Theorem of Operational Task Assignment

We will investigate the validity condition of the organizational level management structure on the IPP.

The target of this chapter is a BDP interpretable organization introduced in Chapter 3. It is defined as:

Definition 3.7 BDP interpretable organization

An organization is called BDP interpretable if there exists an organizational culture BDP:$\underline{D} \rightarrow \wp(Y)$ such that its global goal and operational level goals are always compatible with respect to the BDP, i.e.,

$$BDP(D(\underline{G})) = Opt(G_x)$$

where

$$D(\underline{G}) = <Y, N, g^h>$$

and

$$\underline{D} = \text{class of } D(\underline{G}).$$

The team decision theory of Sec. 8.2 is based on a solution concept of PBPS. The following concept of Nash solution is a generalization of PBPS. The Nash solution is the basis of the main theorem.

Definition 8.2 Nash solution of operational level

$m^* = (m_1^*, .., m_n^*) \in M$ is called *Nash solution* of $(P, \underline{G})$ where $\underline{G} = (G_1, .., G_n)$ if and only if for any i and for any m_i

$$G_i(\Pi_i P(m_i, m_{-i}^*)) \leqq G_i(\Pi_i P(m^*))$$

or

$$G_i(P_i(m_i, m_{-i}^*)) \leqq G_i(P_i(m^*)).$$

Let

$$N(P, \underline{G}) \subset M$$

be the set of Nash solutions of $(P,\underline{G})$. Notice that Nash solution is defined independent of the global goal. It describes an optimum behavior of the operational level.

The following is a basic fact about the IPP.

Proposition 8.3 [2]

The organizational level management structure of Def. 8.1 *satisfies the following:*

$$\textit{If } g_0(r^*) = \text{True, } \textit{then } H^m(m^*(r^*)) \textit{ is a Nash solution of } (P,\underline{G}).$$

The proof is given in Appendix.

The above proposition suggests that the validity situation will be realized when a Nash solution implies an optimum of the global goal. The latter condition is given by following the implementation theory of economics [15].

We use the following notation for the proof of Prop. 8.4.

Let

$$L(y^*,G_i)=\{y|y\in Y \text{ and } G_i(\Pi_i(y))\leqq G_i(\Pi_i(y^*))\},$$

or $L(y^*,G_i)$ is the set of the outputs which are dominated by y^* with respect to the i-th operational level goal G_i.

Proposition 8.4 [15]

Suppose an organization is a BDP interpretable. Suppose the BDP satisfies the following conditions:

***(i) weak no-veto power**:*

For any $\underline{G}$, y^* and i

$$(\forall j)(j \neq i \rightarrow Y=L(y^*,G_j)) \rightarrow y^*\in BDP(D(\underline{G})).$$

(ii) monotonicity:

$$(\forall \underline{G},\underline{G}',y^*)(y^*\in BDP(D(\underline{G})) \text{ and } (\forall i)(L(y^*,G_i) \subset L(y^*,G_i'))$$

$$\rightarrow y^*\in BDP(D(\underline{G}'))).$$

Then, there exist $M_1,\ldots,M_n$ and $P:M_1 \times\cdots\times M_n \rightarrow Y$ such that

$$BDP(D(\underline{G}))=P(N(P,\underline{G}))$$

or

$$y \in BDP(D(\underline{G})) \leftrightarrow (\exists m^*)(m^* \text{ is a Nash solution for the operational level model of } (P,\underline{G}) \text{ and } y=P(m^*)).$$

The proof is given in Appendix.

The proof of Prop. 8.4 is given as a constructive one, which means that the proof presents a procedure to construct P.

The following is the final result.

Proposition 8.5

A valid organizational level management structure with the IPP can be designed for a BDP interpretable organization if its BDP satisfies the conditions of Prop. 8.4.

The proof is given in Appendix.

APPENDIX OF CHAPTER 8

Proof of Proposition 8.1

This is true because $M=M_1 \times \cdots \times M_n$.

Q.E.D.

Proof of Proposition 8.2

Let

$$f(t)=g_x(m^*+t\delta)$$

where $t \in Re$ and $\delta=(\delta_1,..,\delta_n) \in M=Re^n$ is arbitrary. Then,

$$df(t)/dt=\sum_{i=1}^{n} (\partial g_x/\partial m_i)_{m^*+t\delta} * \delta_i.$$

PBPS solution implies

$$(\partial g_x/\partial m_i)_{m^*}=0$$

for each i. Therefore

$$f'(0)=0$$

holds. Since g_x is assumed as a concave function, f(t) is also a concave function. As a matter fact,

$$\begin{aligned} f(kt+(1-k)t') &= g_x(m+(kt+(1-k)t')\delta) \\ &= g_x(km+(1\text{-}k)m+(kt+(1-k)t')\delta) \\ &= g_x(k(m+t\delta)+(1\text{-}k)(m+t'\delta)) \\ &\geqq kg_x(m+t\delta)+(1\text{-}k)g_x(m+t'\delta) \\ &= kf(t)+(1-k)f(t'). \end{aligned}$$

Consequently, since $f'(0)=0$,

$$f(t) \leqq f(0)$$

holds. This comes from the following simple fact. Due to the above inequality

$$f(t'+k(t-t')) \geqq f(t')+k(f(t)-f(t'))$$

for $k \neq 0$. Then, if we take the limit operation $\lim_{k \to 0+}$, we have

$$f'(t')(t-t') \geqq f(t)-f(t').$$

Let $t'=0$. Since $f'(0)=0$,

$$0 \geqq f(t)-f(0)$$

holds. If $t=1$, we have

$$g_x(m^*+\delta) \leqq g_x(m^*).$$

Q.E.D.

Proof of Proposition 8.3

Suppose $g_0(r^*)$=True. Then, the definition of $g_0(r^*)$ implies that

$$\begin{aligned} r^* &= K(H^m(m^*(r^*))) \\ &= K(m^*(r^*)) \end{aligned}$$

where $H^m(m^*(r^*))=m^*(r^*)$. That is,

$$\begin{aligned} r_i^* &= K_i^{'}(m^*(r^*)) \\ &= m_{-i}^*(r^*). \end{aligned} \qquad \text{(A)}$$

The definition of $m^*(r^*)$ implies that

$$g_i(r_i^*)(m_i) \leqq g_i(r_i^*)(m_i^*(r_i^*))$$

or

$$g_i(m_i, r_i^*) \leqq g_i(m_i^*(r_i^*), r_i^*)$$

for any m_i. Then (A) implies that

$$G_i(P_i(m_i, m^*_{-i}(r^*))) = g_i(m_i, m^*_{-i}(r^*)) \leqq g_i(m^*_i(r^*_i), m^*_{-i}(r^*)) = G_i(P_i(m^*(r^*))).$$

Consequently, $m^*(r^*)=H^m(m^*(r^*))$ is a Nash solution.
Q.E.D.

Proof of Proposition 8.4
Let

$$M_i=\{\underline{G}\} \times Y \times N$$

for any i where $N=\{0,1,2,..\}$ and $\{\underline{G}\}$ is the class of $\underline{G}$.

Let $P:M_1 \times \cdots \times M_n \rightarrow Y$ be defined in the following way:

Case 1:

Suppose there exist i and $(\underline{G},y,k)$ such that

$$(\forall j \neq i)(m_j=(\underline{G},y,k)) \text{ and } y \in BDP(D(\underline{G}))$$

holds. Then

$$P(m_1,\ldots,m_n) = \begin{cases} y^i & \text{if } y^i \in L(y, G_i) \\ y & \text{otherwise} \end{cases}$$

where $m_i=(\underline{G}^i,y^i,k^i)$ and $\underline{G}=(G_1,..,G_n)$.

Case 2:

Suppose Case 1 does not hold. If k^i of $m_i=(\underline{G}^i,y^i,k^i)$ is the highest integer among $(m_1,..,m_n)$, let

$$P(m_1,..,m_n)=y^i.$$

If there are more than one highest integers, the smallest i is selected among them.

First we will show

$$BDP(D(\underline{G})) \subset P(N(P,\underline{G})).$$

Let $y^* \in BDP(D(\underline{G}))$ be arbitrary. Let $m^*=(m^*_1,\ldots,m^*_n)$ such that $m^*_i=(\underline{G},y^*,0)$ for every i. Then, Case 1 holds and

$$P(m^*)=y^*.$$

We will show $m^* \in N(P,\underline{G})$. Let i and $m_i=(\underline{G}',y',k')$ be arbitrary. Then, since Case 1 is applicable to (m_i, m^*_{-i}) again, there are two cases for $P(m_i, m^*_{-i})$.

Case A:

If $y' \in L(y^*,G_i)$, $P(m_i, m^*_{-i})=y'$. Since $y' \in L(y^*,G_i)$ implies $G_i(\Pi_i y') \leqq G_i(\Pi_i y^*)$, we have

$$\begin{aligned} G_i(\Pi_i P(m_i, m^*_{-i})) &= G_i(\Pi_i y') \\ &\leqq G_i(\Pi_i y^*) \\ &= G_i(\Pi_i P(m^*)). \end{aligned}$$

Case B:

If $y' \notin L(y^*,G_i)$, $P(m_i, m^*_{-i})=y^*$. Then

$$G_i(\Pi_i P(m_i, m^*_{-i}))=G_i(\Pi_i P(m^*)).$$

Consequently, $m^* \in N(P,\underline{G})$ or $y^*=P(m^*) \in P(N(P,\underline{G}))$.

Next, we will show

$$P(N(P,\underline{G})) \subset BDP(D(\underline{G})).$$

Let $y^* \in P(N(P,\underline{G}))$ be arbitrary. Then

$$y^*=P(m^*) \text{ and } m^* \in N(P,\underline{G}).$$

There are three cases for the proof of $P(N(P,\underline{G})) \subset BDP(D(\underline{G}))$.

Case C:

Suppose there exists $(\underline{G}',y,k)$ such that

$$(\forall i)(m_i^* =(\underline{G}',y,k)) \text{ and } y \in BDP(D(\underline{G}')).$$

In this case, Case 1 is applicable so that

$$y^*=P(m^*)=y.$$

We will show that

$$L(y^*,G_i') \subset L(y^*,G_i)$$

for any i. Let $y' \in L(y^*, G_i')$ be arbitrary. Let $m_i = (\underline{G}', y', k')$. Then, Case 1 is applicable to (m_i, m^*_{-i}) and $P(m_i, m^*_{-i}) = y'$ (because $y' \in L(y^*, G_i')$). Since $m^* \in N(P, \underline{G})$,

$$G_i(\Pi_i P(m_i, m^*_{-i})) \leqq G_i(\Pi_i P(m^*))$$

or

$$G_i(\Pi_i y') \leqq G_i(\Pi_i y^*).$$

Therefore,

$$y' \in L(y^*, G_i).$$

Since $L(y^*, G_i') \subset L(y^*, G_i)$ for any i and since $y^* \in BDP(D(\underline{G}'))$ (because $y^* = y$), the monotonic property of BDP implies $y^* \in BDP(D(\underline{G}))$ which is the desired result.

Case D:

Suppose there exists $(\underline{G}', y, k)$ such that

$$(\forall i)(\ m_i^* = (\underline{G}', y, k)) \text{ and } y \notin BDP(D(\underline{G}')).$$

In this case Case 2 is applicable to m^* such that

$$y^* = P(m^*) = y.$$

We will show

$$L(y^*, G_i) = Y \text{ for all i.}$$

Let i and $y' \in Y$ be arbitrary. Let

$$m_i = (\underline{G}', y', k')$$

where $k' > k$. Then, Case 2 is applicable to (m_i, m^*_{-i}) so that $P(m_i, m^*_{-i}) = y'$.

Since $m^* \in N(P, \underline{G})$,

$$G_i(\Pi_i P(m_i, m^*_{-i})) \leqq G_i(\Pi_i P(m^*))$$

or

$$G_i(\Pi_i y') \leqq G_i(\Pi_i y^*).$$

Hence, $y' \in L(y^*, G_i)$. Consequently, $L(y^*, G_i) = Y$ for all i. Then, the weak no-veto power implies that

$$y^* \in BDP(D(\underline{G}))$$

which is the desired result.

Case E:

Suppose that it is not true that every component of m^* is equal to each other. Let us assume $m_1^* \neq m_2^*$ without losing generality.

Let i be arbitrary such that $i \neq 1$ and $i \neq 2$. We will show

$$L(y^*, G_i) = Y.$$

Let $y' \in Y$ be arbitrary. Let us denote

$$m_j \equiv (\underline{G}^j, y^j, k^j) \text{ for all } j \neq i.$$

Let

$$m_i = (\underline{G}, y', k')$$

where $k' > k^j$ $(j \neq i)$. Then, Case 2 is applicable to (m_i, m_{-i}^*) so that $P(m_i, m_{-i}^*) = y'$.

Since $m^* \in N(P, \underline{G})$,

$$G_i(\Pi_i P(m_i, m_{-i}^*)) \leqq G_i(\Pi_i P(m^*))$$

or

$$G_i(\Pi_i y') \leqq G_i(\Pi_i y^*).$$

Hence,

$$y' \in L(y^*, G_i).$$

Consequently, $L(y^*, G_i) = Y$ for all i where $i \neq 1,2$.

When $i \neq 1,2$, then $m_i^* \neq m_1^*$ or $m_i^* \neq m_2^*$ should hold. Let us assume $m_i \neq m_1$. Then, the above argument is applicable to (m_1^*, m_i^*) such that we have $L(y^*, G_2) = Y$. (2 is substituted for i in the above argument).

Then, the weak no-veto power implies that

$$y^* \in BDP(D(\underline{G}))$$

which is the desired result.

In summary, we have $y^* \in BDP(D(\underline{G}))$ for every case.

Q.E.D.

Proof of Proposition 8.5

Let $(\{Y_i\}_i, \underline{G}, G_x)$ be the specification of a BDP interpretable organization, i.e., $BDP(D(\underline{G})) = Opt(G_x)$. Since BDP satisfies the conditions of Prop. 8.4, we have $\{M_1,..,M_n\}$ and $P: M_1 \times \cdots \times M_n \rightarrow Y$ for which

$$P(N(P,\underline{G}))=BDP(D(\underline{G}))=Opt(G_x)$$

holds.

Prop. 8.3 asserts that

$$g_0(r)=True \rightarrow H^m(m^*(r))=m^*(r) \in N(P,\underline{G}).$$

Consequently,

$$y^*(r)=P(m^*(r)) \in Opt(G_x).$$

Therefore, the validity condition is satisfied.

Q.E.D.

CHAPTER 9 NORMATIVE BEHAVIOR MODEL OF OPERATIONAL LEVEL

PROCESS MODEL FOR ADAPTATION

As a preliminary to Chapter 10 this chapter will examine a management of the operational level in an ideal form. The model of this chapter is not a real one. It provides a normative image of the primary activity of an organization, based on which the institutional level plans its adaptive activity. The image is a process model of the institutional management.

An organization of the simplest form consists of the three levels—process level, operational level and management level—and an organizational goal. The simplest structure is shown in Fig. 9.1 of Sec. 9.1. The institutional level and the organizational level constitute the management level. The total or external behavior of an organization in this representation is an input-output system whose input is a management information input to the management level and whose output is a product from the process level.

The institutional level receives the management information input from environment and combines it with the organizational goal to produce a global goal. This is an adaptive activity of the institutional level, which will be discussed in Chapter 10. The activity is supposed to use a simple normative view of the primary activity of the operational level although real management is carried out through a complicated coordination structure. The global goal is a control variable of institutional level to manage the entire organization.

This chapter assumes the departmentalized structure for the operational level as before. Each sub-process is also controlled by the corresponding operational decision maker. The difference between the normative model in this chapter and the previous ones is that a sub-unit of the operational level in this chapter is modeled as an input/output system whose input is a directive from the management level. The sub-unit executes the directive exactly as it is designed and yields a product of the organization as its output.

No decision making activity is assumed for the operational level in the present model. According to Sec. 2.4 a bureaucratic model is then applied to the operational level. The operational units are faithful bureaucrats. Certainly, this faithfulness is a basis for adaptation planning of the institutional level.

In the normative model, the organizational level management is supposed to allocate subtasks to the operational units in accordance to the planned directive so that the directive can be realized as efficiently as possible. The allocation should be done subject to management resource constraints.

This is an idealized management situation. The main question of this chapter is, then: How to allocate subtasks for the operational level to implement the directive of the management level satisfactorily.

Most of the theoretical results of this chapter come from Ref. [16] .

9.1 Normative Operational Level Model

Fig. 9.1 shows the organization model used in this chapter.

The operational level consists of n input output subsystems. They are not recognized as goal seeking systems. The subsystems are denoted by $Op_1,..,Op_n$. Op_i will be called i-th operational unit. The i-th operational unit consists of the i-th sub-process and the i-th operator. Since a decision maker of the operational level is assumed to function as an input/output system, it will be called operator in this chapter.

The input to Op_i is a task assignment or a directive v_i from the management level. The output of Op_i is denoted by y_i.

In Chapter 4 we discussed the structure of an organization ignoring personal differences among the organization members. In that chapter, since it was assumed that every member have the same capability and can operate without interaction, the production capacity of an organization was assumed to be determined by the number of operators who have direct control over the process.

It is of course not true that every member of an organization has the same capability. It is usual that a person is talented in some field while he is weak in another field. When the goal seeking model is used for a member of an organization, the model can differentiate members by their goals, motivation and cognition processes. In this chapter since an operator is modeled as an input/output system, some other features are used to represent personal differences.

In this chapter every Op_i is assumed to have the same input/output behavior. In this sense there is no difference among the operational units. But it is assumed that the processing time required to fulfill a given task (transform an input into an output) depends on an operational unit as well as

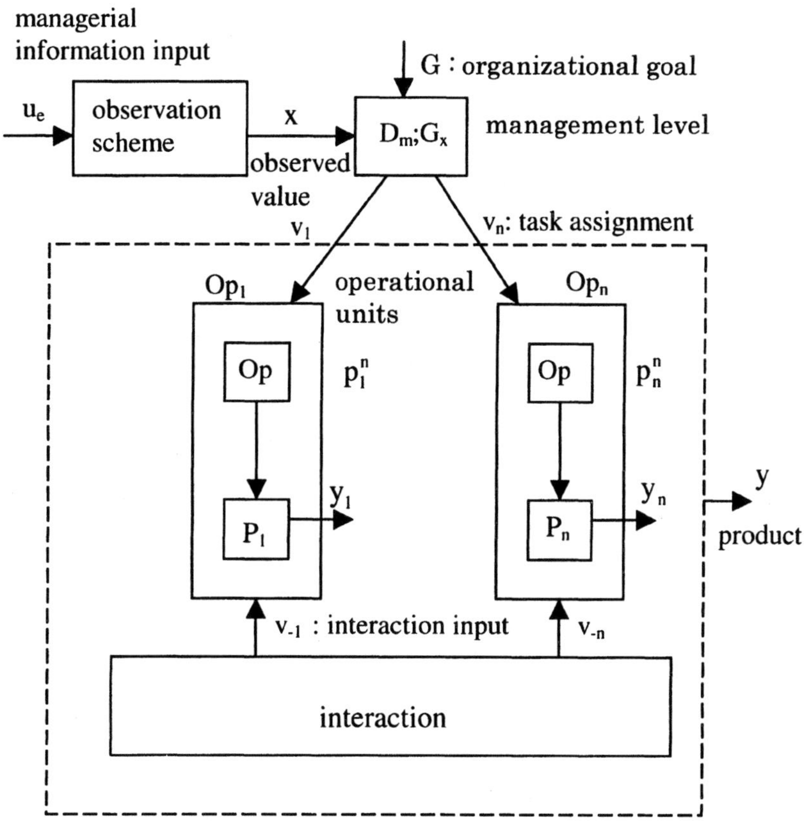

Figure 9.1 Normative organization model

on the type of the task. Efficiency is different among the operators. Then the tasks must be assigned taking into account efficiency of the operational units. An efficient or feasible task assignment among the operational units is a management problem of the organizational level in this chapter. The organizational management which performs this assignment will be called an *idealized organizational level management.*

It must be noticed that an organization must satisfy the validity condition. However, the validity is not in general sufficient for management although it must be taken into account for successful and timely management. This chapter is concerned with efficiency.

The input of the management unit D_m of Fig. 9.1, which consists of the institutional and organizational levels, is an observed value x of an external input u_e which is assumed to be a random variable. The input is not observed continuously but periodically or the rate of change u_e is assumed to be quite slow compared with the dynamics of the operational level so that an appreciable change of u_e can be observed only after some period of time has passed. The period will be equalized to the coordination period T_{coord} introduced in Chapter 5. The operational task derived from (observed) input is then required to be processed within T_{coord}. This requirement will be used as a criterion of efficiency. It will be relaxed in Chapter 11.

The output variable $(v_1,..,v_n)$ of D_m, —a directive of task assignment—is an input to the operational level. The task assignment of D_m will be assumed to be executed in a probabilistic way. This assumption can make it possible to avoid the notorious combinatorial optimization problem. The assignment scheme will be called *probabilistic assignment protocol* [16].

9.2 Formulation of Normative Operational Unit and Optimal Directive

This section will formalize the system shown in Fig. 9.1.

Definition 9.1 Operational units of normative model
Let

$$N=\{1,2,..,n\}$$

be the index set of the operational units.

Let

$$V_i = \text{set of directives to the i-th operational unit,}$$

is assumed finite, i.e.,

$$V_i=\{v_i^1,v_i^2,\ldots\}=\{v_i^j|j\in J_i\}$$

where J_i is finite. $v_i^j \in V_i$ is a j-th directive from D_m to the i-th operational unit.

Let the total input to the operational level be:

$$V = V_1 \times V_2 \times \cdots \times V_n.$$

Notice that V is a finite set, whose element will be denoted as

$$v^j=(v_1^{j1},\ldots,v_n^{jn}).$$

The finite index set of j will be denoted by J.

Let

Y_i= output set of the i-th operational unit,

which is also assumed finite.

Let the total output of the operational level or the organization be:

$$Y = Y_1 \times Y_2 \times \cdots \times Y_n.$$

Notice that Y is a finite set whose element will be denoted as

$$y^j=(y_1^{j1},\ldots,y_n^{jn}).$$

Let the imaginary (normative) input/output relation of the i-th operational unit be:

$$P_i^n : V \rightarrow Y_i.$$

It is assumed that in general there are interactions among the sub-processes. Let an interaction be represented by v_{-i} where

$$v_{-i}=(v_1,..,v_{i-1},v_{i+1},..,v_n).$$

Let the time, which Op_i must spend to process v_i^j, be given by a function

$$\tau_i:V_i \rightarrow Re .$$

That is, when v_i^j is assigned to Op_i, Op_i spends $\tau_i(v_i^j)$ to fulfil it.

Then, the specification of the i-th *operational unit* is given by

$$\text{i-th operational unit}=<V_i,V_{-i},Y_i,P_i^n, \tau_i >.$$

Finally, let the normative overall input/output relation of the operational level be:

$$P^n : V \rightarrow Y$$

where

$$P^n(v)=(P_1^n(v),..,P_n^n(v)).$$

An obvious and simple interpretation of the task assignment directive is an optimal outcome y^* of $E_{u_e|x} G(u_e,y)$. The directive is a command that the operational level should produce y^* within an assigned time limit. This assignment, however, may not be feasible.

An observation of the external information input is done at the beginning of each interval of T_{coord}.

Definition 9.2 External managerial information input
Let

U_e=set of values of the external information input,

(= the set of states of the external environment).
Let

X=the set of observed values of the external information input.

U_e and X are assumed finite. Their elements will be denoted by $u_{ek} \in U_e$ (k=1,2,..) and $x_j \in X$ (j=1,2,..), respectively. u_{ek} and x_j may be time functions.

It is also assumed that the observed value is an independent random variable whose probability model is given by:

probability model of the observed input=$<X, \wp(X), Pr_x>$

where $Pr_x: \wp(X) \rightarrow [0,1]$ is a probability function.

Furthermore, the external input is assumed modeled by the following conditional probability:

$Pr(u_e|x)$.

The organizational goal G is

$G : U_e \times Y \rightarrow Re.$

For an observed value x_j let an optimal outcome y^k be specified by the following relation:

$(\forall\ y \in Y)(E_{u_e|x_j} G(u_e,y) \leqq E_{u_e|x_j} G(u_e,y^k)).$

Then, let $\delta^h: X \rightarrow Y$ be

$$y^k=\delta^h(x_j) \rightarrow (\forall\ y \in Y)(\ E_{u_e|x_j} G(u_e,y) \leqq E_{u_e|x_j} G(u_e, \delta^h(x_j))).$$

$\delta^h()$ has a following desirable property.

Proposition 9.1
Let

$$max_y E_{u_e} G(u_e,y)= E_{u_e} (u_e,y^*).$$

Then,

$$E_x E_{u_e|x} G(u_e, \delta^h(x)) \geqq E_{u_e} (u_e,y^*).$$

The proof is given in Appendix.

y^* is the most desirable outcome expected by G. But the proposition says that the outcome $\delta^h(x)$ derived adaptively depending on the observed value x produces on average a better performance than y^*. This follows from the fact that $\delta^h(x)$ is selected as the most desirable from a class of mixed strategies while y^* is selected from a class of pure strategies. The class of mixed strategies covers that of pure strategies.

The next proposition shows how an optimum directive is determined in relation to $\delta^h(x)$.

Proposition 9.2
Let $\delta^*:X \rightarrow V$ *be an optimal decision function as defined in Prop. 2.1, i.e.,*

$$max_{\delta} E_x E_{u_e|x} G(u_e,P^n(\delta(x)))=E_x E_{u_e|x} G(u_e,P^n(\delta^*(x))).$$

Suppose P^n *is an onto function. Then the following holds:*

$$E_{u_e|x} G(u_e, \delta^h(x))= E_{u_e|x} G(u_e,P^n(\ \delta^*(x))).$$

The proof is given in Appendix.

The fact that P^n is an onto function can be interpreted as that the variety of the management level can compete with the primary activity. We always assume this variety matching condition in this book.

The above proposition says that an optimum outcome $y^k=\delta^h(x)$ is given by

$$y^k=P^n(\delta^*(x))$$

or an optimal directive is given by $\delta^*(x)$ for an observed input x.

Definition 9.3 Optimum directive
Let $\delta^*()$ be the optimum decision function of Prop. 9.2. Then, $v^*=\delta^*(x)$ is called optimum directive of x.

The relations among δ^h, δ^* and P^n are given by the commutative diagram of Fig. 9.2.

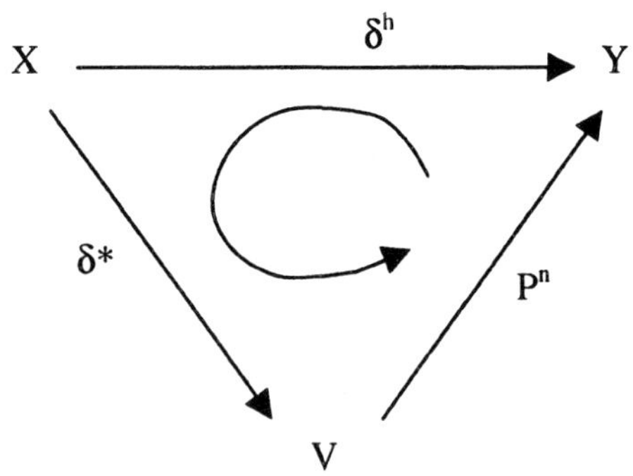

Figure 9.2 Factorization of dh

Notice that this fact holds because the variety matching is satisfied and G is a function of y but not of v.

For the sake of notational convenience let us write y^j for an optimum outcome with respect to x_j, i.e.,

$$\max_y E_{u_c|x_j} G(u_e,y)= E_{u_c|x_j} G(u_e,y^j)$$

or

$$y^j=\delta^h(x_j).$$

9.3 Management Structure of Normative Model

We will now investigate feasibility of $\delta^*()$ using the concept of a probabilistic assignment protocol of Ref [16].

First, the following three simplifications will be made about the normative task model.

(1) Let π be a permutation on $\{1,..,n\}$ or $\pi:\{1,..,n\} \rightarrow \{1,..,n\}$ and π is a one to one correspondence. Then we will denote a permutation on $y=(y_1,..,y_n)$ as

$$\pi(y)=(y_{k1},..,y_{kn})$$

where $i=\pi(ki)$. The same notation will be used for $v=(v_1,..,v_n) \in V$. Then the following relation is assumed for G: For any u_e and for any permutation π

$$G(u_e,y)=G(u_e,\pi(y)).$$

(2) The operational units are assumed isomorphic to each other, or P^n satisfies the following relation:

$$\pi(P^n(v))=P^n(\pi(v)) \text{ for any } \pi \text{ and } v.$$

(3) P^n is assumed to be an onto function, i.e., for each optimum outcome y^j there exists $v \in V$ such that $y^j = P^n(v)$, or the variety matching holds between the management level and the primary activity.

Let

$$\begin{aligned} Perm &= \text{the set of permutations} \\ &= \{\pi \mid \pi : N \rightarrow N\}. \end{aligned}$$

Let x_j be an observed input. x_j is fixed in the subsequent discussions.
Let $y^j= P^n(\delta^*(x_j))= \delta^h(x_j)$ be the optimum outcome for x_j, i.e.,

$$(\forall y)(\ E_{u_e|x_j} G(u_e,y) \leqq E_{u_e|x_j} G(u_e,y^j))$$

where

$$y^j=(y^j{}_1,..,y^j{}_n)$$

and

$v^j=(v^j_1,..,v^j_n)$.

Since P^n is assumed an onto function, for each x_j, $v^j=\delta^*(x_j)$ is an optimal directive in the sense of Prop. 9.2 and hence D_m should assign a task $v^j=(v^j_1,..,v^j_n)$ to the operational level if the efficiency factor τ_i is ignored. There is no choice for D_m if x_j is given. But since it is assumed that $G(u_e,y^j)=G(u_e,\pi(y^j))$ for a permutation π, $\pi(v^j)$ is another candidate of the optimum task assignment. (Notice that $\pi(y^j)=\pi P^n(v^j)=P^n\pi(v^j)$.) That is, since it is desirable that D_m selects a π which produces the most efficient task assignment $\pi(v^j)$, π is a decision variable of D_m in the present formulation.

Let us assume that selection of π or the task assignment is done in a probabilistic way. Suppose the selection obeys the following probability model:

$$\underline{Pr}=\{Pr(\pi|x_j)\mid\pi\in\text{Perm}\}.$$

That is, the assignment of $\pi(v^j)$ (or $\pi(y^j)$) is made with the conditional probability $Pr(\pi|x_j)$. Let $\underline{Pr}$ be called *probabilistic assignment protocol* [16]. Then, $\underline{Pr}$ is the real decision variable of D_m. The decision of D_m is a mixed strategy.

When a permutation π is selected for the observed value x_j, the i-th operational unit needs the following time to accomplish the assignment:

$$\tau_i(v^j_{ki})$$

where $ki=\pi^{-1}(i)$. The above value gives a measure of the efficiency of the normative management.

Using the conditional probability $Pr(\pi|x_j)$, the processing time that the i-th operational unit should spend during T_{coord} is on average given by

$$\underline{\tau_i}(\underline{Pr})=\sum_k Pr_i(v^j_k\mid x_j)\tau_i(v^j_k)$$

where $Pr_i(v^j_k|x_j)$ is the conditional probability that the directive v^j_k is assigned to the i-th operational unit when x_j is observed, which is given by

$$Pr_i(v^j_k|x_j)=\sum_\pi\{Pr(\pi\mid x_j)\mid\pi(k)=i\}.$$

$\underline{\tau_i}(\underline{Pr})$ provides an efficient measure of $\underline{Pr}$ of the i-th unit. Then, let the goal g_m of D_m be given by the following definition.

Definition 9.4 Decision goal of normative management
Let the decision goal be $g_m:\{\underline{Pr}\}\rightarrow\{\text{True},\text{False}\}$ be

$$g_m(\underline{Pr})=\text{True} \leftrightarrow \max_i \tau_i(\underline{Pr}) < T_{coord}.$$

If $g_m(\underline{Pr})$=True, $\underline{Pr}$ is said to be *feasible.*

If the above condition is not satisfied, a backlog would be accumulated and the operation of the organization would be broken or a normal management might not be maintained.
Then, we have

Definition 9.5 Management structure of Normative model
A *management structure of normative model* is given by

$$\text{management structure of normative model}=<\{\underline{Pr}\},g_m>.$$

Fig. 9.3 shows the total structure of the normative model.

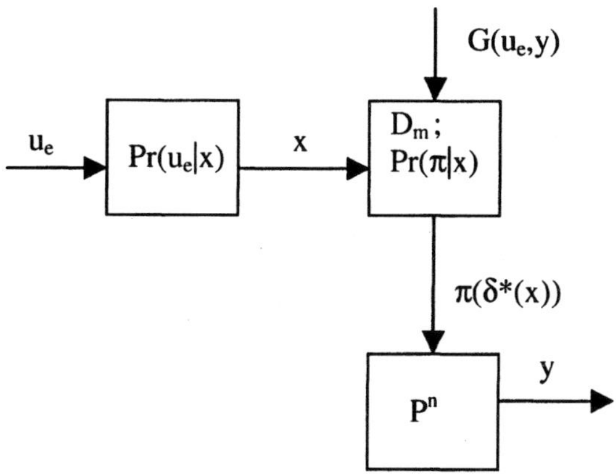

Figure 9.3 Management in normative model

9.4 Feasibility of Normative Management

Using the above formulation let us investigate feasibility of a normative management.

Definition 9.6 No overload condition
Let Δ be a real number which satisfies

$\Delta < T_{coord}$.

$\underline{\tau_i}(\underline{Pr}) < \Delta$ for every i, $\underline{Pr}$ is said to satisfy the no overload condition. Conversely, $\underline{\tau_i}(\underline{Pr}) \geqq \Delta$ for some i, we say that $\underline{Pr}$ breaks the overload condition.

Δ is an upper limit of the average working time for an operator. This condition is more restrictive than the feasibility condition.

The following is a conceptual result.

Proposition 9.3 [16]

Suppose there is an integer n *such that*

(1) there exists a function $k^*:N \times J \rightarrow N$ *such that* $k^*(-,j):N \rightarrow N$ *is a one to one correspondence;*

(2) for every $(i,j) \in N \times J$ *the processing time* τ_i *has the following property*

$$\tau_i(v^j_k) < \Delta$$

for $k=k^*(i,j) \in N$.

Then there is an assignment protocol where overload on the operational level can be avoided for the n-th unit *where $N=\{1,\ldots,n\}$.*

The proof is given in Appendix.

Definition 9.7 Specialist

An operational unit i is called specialist if, for any directive v^j, there is a sub-directive $v^j_{i_j}$ of v^j such that the unit is the only one which can process within Δ. $v^j_{i_j}$ is called field of specialization of the operational unit, i.

Corollary 9.1 [16]

Suppose every member of the operational level is a specialist and their fields are different from each other. Then there is an assignment protocol such that the overload can be avoided.

The proof is given in Appendix.

The following is a basic characterization of the feasibility.

Proposition 9.4 [16]

A feasible assignment protocol with respect to Δ *exists for the operational level of the size* n *only if there exists a linear programming problem whose variables are assignment probabilities* $\Pr(\pi|x_j)$ *such that the problem has a feasible solution. Conversely, if the problem has a feasible solution, a feasible assignment protocol with respect to* Δ *exists for the operational level.*

The proof is given in Appendix.

APPENDIX OF CHAPTER 9

Proof of Proposition 9.1

The same argument as used in Prop. 2.1 implies that

$$\max_{\delta} E_x\ E_{u_e|x}\ G(u_e, \delta(x))=E_x\ E_{u_e|x}\ G(u_e, \delta^h(x)).$$

Since a constant function $\underline{\delta}(x)=y^*$ is a candidate of δ,

$$\begin{aligned} &E_x E_{u_e|x} G(u_e, \delta^h(x)) \geqq E_x E_{u_e|x} G(u_e, \underline{\delta}(x)) \\ &= E_x E_{x|u_e} G(u_e, y^*) \\ &= E_{u_e} E_{u_e|x} G(u_e, y^*) \\ &= E_{u_e} G(u_e, y^*) \ \text{(because } y^* \text{ is independent of x).} \end{aligned}$$

Q.E.D.

Proof of Proposition 9.2

Suppose an observed input x in X is given. Let$\delta^*(x)=v^*$ and $y^*=P^n(v^*)$. Then, the definition of δ^* implies that

$$\max_v\ E_{u_e|x}\ G(u_e, P^n(v))=\ E_{u_e|x}\ G(u, P^n(v^*))=E_{u_e|x}\ G(u_e, y^*).$$

Let

$$\max_y E_{u_e|x}\ G(u_e, y)=\ E_{u_e|x}\ G(u_e, \underline{y}).$$

Then, the following holds:

$$E_{u_e|x}\ G(u_e, \underline{y}) \geqq\ E_{u_e|x}\ G(u_e, y^*).$$

We will show

$$E_{u_e|x}\ G(u_e, \underline{y}) \leqq\ E_{u_e|x}\ G(u_e, y^*).$$

Since P^n is an onto by assumption, there exits $\underline{v} \in V$ such that

$$\underline{y}=P^n(\underline{v}).$$

Hence, the definition of v^* implies

$$E_{u_e|x}\, G(u_e,y^*) = E_{u_e|x}\, G(u_e,P^n(v^*)) \geqq E_{u_e|x}\, G(u_e,P^n(\underline{v})) = E_{u_e|x}\, G(u_e,\underline{y}).$$

Consequently,

$$\begin{aligned}
&\max_y E_{u_e|x} G(u_e, y) = E_{u_e|x} G(u_e, \underline{y}) = E_{u_e|x} G(u_e, y^*) \\
&= E_{u_e|x} G(u_e, P^n(v^*)) \\
&= \max_v E_{u_e|x} G(u_e, P^n(v)).
\end{aligned}$$

Q.E.D.

Proof of Proposition 9.3
Suppose i and j are fixed. Let $\underline{Pr}$ be:

$$\Pr(\pi \mid x_j) = \begin{cases} 1 \text{ if } \pi^{-1} = k^*(-, j) \\ 0 \text{ otherwise} \end{cases}$$

Then

$$\Pr_i(v_k^j \mid x_j) = \begin{cases} 1 \text{ if } k = k^*(i, j) \\ 0 \text{ otherwise} \end{cases}$$

Consequently, for each i

$$\underline{\tau_i}(\underline{Pr}) = \sum_k \Pr_i(v_k^j \mid x_j)\tau_i(v_k^j)$$

$$= \tau_i(v^j_{kj})$$

(where $kj = k^*(i,j)$)

$$< \Delta.$$

Q.E.D.

Proof of Corollary 9.1

Let $k^*(i,j)$ of Prop. 9.3 be: $i_j=k^*(i,j)$ where i_j is the specialization field of the i-th member for v^j. Then, following the same argument of the proof of Prop. 9.3 we have

$$\underline{\tau_j(\text{Pr})} = \tau_i(v^j_{i_j})$$

$$< \Delta.$$

Q.E.D.

Proof of Proposition 9.4

Let

$$X_{j\pi} \equiv \Pr(\pi | x_j).$$

Then

$$\Pr_i(v^j_k | x_j) = \sum \{X_{j\pi} \mid \pi(k) = i\}.$$

In order to show that $\Pr_i(v^j_k|x_j)$ is a linear combination of $\{X_{j\pi}\}$ let us consider the case where J={1,2} and N={1,2,3 } as an example.

In this case Perm={ $\pi_1,.., \pi_6$} where

$$\pi_1:(1,2,3) \rightarrow (1,2,3)$$

$$\pi_2:(1,2,3) \rightarrow (1,3,2)$$

$$\pi_3:(1,2,3) \rightarrow (2,1,3)$$

$$\pi_4:(1,2,3) \rightarrow (2,3,1)$$

$$\pi_5:(1,2,3) \rightarrow (3,1,2)$$

$$\pi_6:(1,2,3) \rightarrow (3,2,1)$$

where, for instance, $\pi_2(1)=1, \pi_2(2)=3, \pi_2(3)=2$.

Let us denote X_{jl} for $\pi=\pi_l$.

Suppose i=2 . Then

$$Pr_2(v^j_k|x_j)=\sum\{X_{j\pi}|\ \pi(k)=2\}.$$

If k=1, since

$$\pi_3(1)=2$$

and

$$\pi_4(1)=2,$$

we have

$$Pr_2(v^j_1|x_j)=\sum\{X_{j\pi}|\ \pi(1)=2\}$$

$$=X_{j3}+X_{j4}.$$

In the similar way we have

$$Pr_2(v^j_2|x_j)=X_{j1}+X_{j6},$$

$$Pr_2(v^j_3|x_j)=X_{j2}+X_{j5}.$$

Consequently, in general, we have the following representation:

$$Pr_i(v^j_k|x_j)=\sum_\pi \alpha(i, k, \pi)X_{j\pi}$$

where $\alpha(i,k,\pi)$ is an appropriate coefficient. For instance,

$$\alpha(2,1,1)=0$$

and

$$\alpha(2,1,3)=1.$$

Then, due to the definition of $\underline{\tau_j}(\underline{Pr})$

$$\underline{\tau}_i(\mathrm{Pr}) = \sum_k \left(\sum_\pi \alpha(i,k,\pi) X_{j\pi}\right) \tau_i(v_k^j)$$
$$= \sum_\pi \left(\sum_k \alpha(i,k,\pi) \tau_i(v_k^j)\right) X_{j\pi}$$
$$= \sum_\pi \beta(i,j,\pi) X_{j\pi}$$

where

$$\beta(i,j,\pi) = \sum_k \alpha(i,k,\pi) \tau_i(v_k^j).$$

Let a linear programming problem be:

$$\sum_\pi \gamma_{j\pi} X_{j\pi} \to \min$$

subject to

$$\sum_\pi \beta(i,j,\pi) X_{j\pi} \leqq \Delta,$$

$$\sum_\pi X_{j\pi} = 1 ,$$

$$X_{j\pi} \geqq 0.$$

Then, if a feasible protocol exists, the above linear programming problem has a feasible solution because $\underline{\tau}_i(\underline{\mathrm{Pr}}) \leqq \Delta$ for any i.

Conversely, if the above problem has a feasible solution, a feasible protocol exists because $\underline{\tau}_i(\underline{\mathrm{Pr}}) = \Sigma_\pi \beta(i,j,\pi) X_{j\pi} \leqq \Delta$.

Q.E.D.

CHAPTER 10 INSTITUTIONAL LEVEL MANAGEMENT

ADAPTATION ACTIVITY

This chapter will discuss an adaptive behavior of an organization performed by the institutional level management. In order to get sharp results, the adaptation of this chapter will be particularized as an activity to adapt to a structural change of the external management information input. The adaptation is performed by adjusting a decision parameter of the institutional level. The ultimate outcome is a global goal which is an operationalization of the organizational goal adapting to the external input change. The goal of the adaptation is to sustain a satisfactory activity state (a set point of the institutional level management) of the organization. It should be noticed then that this adaptation concept presumes that the organization is in an acceptable state in an appropriately defined sense before an adaptation activity is requested.

In this context the most basic questions certainly are: What is the structural change of the external input means, what does a decision parameter of the institutional level means, and what is a satisfactory condition of the organization performance means.

In Sec. 10.1 the external input structure and the institutional management structure will be introduced in a formal way. Subsequent arguments about adaptation will then be developed. Since adaptation is concerned in this chapter in reference to the overall performance of the organization, not in reference to the detailed sub-models, which has been a subject of chapters until Chapter 8, the normative model of Chapter 9 will be used as a comprehensive process model (primary activity) in this chapter. The structural change of the external input will be represented by the structural change of its probabilistic model.

Provided with a probabilistic model of the external input and the normative model, the organizational goal can determine an optimal directive for the operational level for each observed value of the external input. The correspondence δ^* from the observed value to the optimal directive is an ideal decision function of management of the primary activity. Realization and maintenance of the ideal decision function δ^* against a structural change of the external input is then a goal of the institutional level management. δ^* is generated by a function ϕ, the adaptation goal generator which will be represented as a composite of the probability model of the observation scheme, the normative model and the organizational goal. This is the most important function of this chapter.

Using the goal generator ϕ, this section will introduce two basic concepts: Adaptation goal controllability and ϕ-α consistency. The adaptation goal controllability indicates what capability the management structure of the institutional level should have for satisfactory adaptation. This represents an internal requirement of the level. ϕ-α consistency, on the other hand, specifies a consistency between dynamism α of the structural change of the environment and the adaptation goal generator. This represents an external requirement for the level. Satisfactory adaptation can be achieved only when the internal and external requirements are satisfied.

Sec. 10.2 will introduce the concepts of the internal model principle. It characterizes the adaptive behavior when the adaptation goal controllability and the ϕ-α consistency are satisfied simultaneously. The internal model is an image of the dynamism of the environment. It will be shown that an ideal adaptation corresponding to ϕ can be maintained if the organization has a proper internal model. It may be intuitively clear that if the property of the environment is known, an ideal adaptation can be implemented. The result of Sec. 10.2 also shows that an optimal adaptation demands existence of an internal model.

The adaptation goal controllability requires a factorization of the goal generator ϕ with respect to the observation mechanism of the structural change of the external input. If the factorization is possible, the remaining factor specifies an ideal decision of the institutional level.

Sec. 10.3 will define the concept of environmental variety and internal variety. The definitions will be made in relation to the adaptation goal generator ϕ and the dynamism α of the environment. It will be shown that a satisfactory formal definition of varieties cannot be obtained simply by using the range of the environmental structure change or that of the internal structure change.

Sec. 10.4 will discuss adaptation by variety matching between the internal and external (environmental) varieties. What is shown is a formal representation of the law of requisite variety that says that if the environmental variety increases, the necessary internal variety must increase proportionally if an organization is to be viable.

10.1 External Input Structure and Institutional Level Management Structure

a. External Input Structure

The external input $u_e \in U_e$ has been assumed as a random variable. Furthermore, a (subjective) probability model is assumed for U_e. U_e is observed by an observation scheme whose output set is X. Then, this chapter assumes that the structural change of the environment is given by the change of its probability distribution. In general let

U_s = parameter set to characterize the structure of the probability model of U_e.

In particular let us assume that the conditional probability $Pr(u_e|x)$ is parameterized by U_s or

$$Pr(u_e|x,u_s)$$

where $x \in X$ is an observed value of U_e and $u_s \in U_s$.

It is assumed that change of the structure is slow and hence the structural change is observed at a discretized time whose interval T_{adapt} is longer than T_{coord} as defined in Chapter 9. T_{coord} corresponds to the behavior change and not to the structural change of the external input process.

Let the dynamics of U_s be modeled as below.

Definition 10.1 Dynamics of external input structure

Let a *dynamics of the external input structure* be

$$U_s = \langle U_s, \alpha \rangle$$

where α is any transformation on U_s,

$$\alpha: U_s \rightarrow U_s.$$

$\langle U_s, \alpha \rangle$ will be called input structure.

Let us consider the following example for the external input:

$$u_e(t) = y_s(t) + n(t)$$

where $u_e(t)$, $y_s(t)$, $n(t) \in Re$ and y_s represents a structural change of u_e while n represents the "noise". Suppose $y_s(t)$ is given by the following difference equation:

$$y_s(k+2)+ay_s(k+1)+by_s(k)=0$$

where the difference is specified by the time interval T_{adapt}. Then, setting $y_s(k+1)=z_s(k)$, the difference equation can be transformed into a state equation

$$\begin{pmatrix} y_s \\ z_s \end{pmatrix} (k+1) = \begin{pmatrix} 0 & 1 \\ -b & -a \end{pmatrix} \begin{pmatrix} y_s \\ z_s \end{pmatrix} (k).$$

Let

$$u_s = \begin{pmatrix} y_s \\ z_s \end{pmatrix} (k) \in Re^2$$

and

$$\alpha = \begin{pmatrix} 0 & 1 \\ -b & -a \end{pmatrix}.$$

Then, the dynamics of u_s is given by $<U_s,\alpha>$ and if the probability model of n(t) is given by $Pr^n(n(t)|x(t))$, the model of u_e is given by

$$Pr^n(u_e-(1,0)u_s|x)$$

which determines $Pr(u_e|x,u_s)$.

As the above example indicates, since u_s is a vector in general, the specification of the external input by $U_s=<U_s,\alpha>$ and $Pr(u_e|x,u_s)$ can cover a wide range of applications.

b. Institutional Level Management Structure

In order to develop a formal theory of the institutional level management we will make the following assumptions

Assumption 1

The institutional level management has a normative image of the behavior of the operational level and its overall process, which is given by

$$P^n: V \rightarrow Y.$$

$v \in V$ represents an imaginary intervention from the management level to the operational level as discussed in Chapter 9. P^n is an image of the primary activity of the organization generated by a "walk around" by the president. The adaptive activity of the institutional level is based on P^n.

Assumption 2

The institutional level management has a function

$$\chi: V \rightarrow \text{hom}(Y, Re)$$

which is called *organization policy*. It transforms an imaginary intervention into a global goal. This is a function performed by the level 5 of the viable system model of Chapter 1. It is also a key factor to link the macroscopic behavior and the microscopic behavior of an organization. hom(Y,Re) $(=\{G_x | G_x: Y \rightarrow Re\})$ represents the set of global goals.

χ is assumed to satisfy the following relation: For any $v \in V$

$$P^n(v) = y^* \rightarrow \max_y \chi(v)(y) = \chi(v)(y^*). \qquad \text{(A)}$$

Since $\chi(v): Y \rightarrow Re$ is a global goal, y^* is an optimal solution of it. Then, the relation says that if y^* is an output of the normative model at v, then it must be an optimal solution of the global goal generated by χ. This is a natural requirement due to the meaning of χ. If Y is a set of real valued vectors, the following is a possible example of χ:

$$\chi(v)(y) = G_x(y) = -(y - P^n(v))^T Q (y - P^n(v))$$

where Q is a non-negative symmetric matrix. χ may be considered as a solution of the property (A).

Assumption 3

The institutional level management activity is represented by a decision function

$$\delta: X \rightarrow V.$$

An optimal decision function $\delta^*: X \rightarrow V$ was defined using the organizational goal G in Chapter 9. That is,

$$\max_\delta E_{u_e} E_{x|u_e} G(u_e, P^n(\delta(x))) = E_{u_e} E_{x|u_e} G(u_e, P^n(\delta^*(x))).$$

δ^* produces the best expected value of the organizational goal. An ideal behavior of the institutional management is equal to δ^*. It is a set point for the institutional management.

As Prop. 2.1 shows, δ^* is characterized by

$$\delta^*(x)=v^* \rightarrow \max_v E_{u_e|x} G(u_e,P^n(v))= E_{u_e|x} G(u_e,P^n(v^*)).$$

The desired behavior of the institutional level management is to keep the decision function optimal for any environmental change.

It should be noticed that $\delta^*:X \rightarrow V$ depends on the conditional probability of U_e on X and since the probability is assumed to change following an environmental change, it depends on change of the parameter $u_s \in U_s$. When the dependence is to be expressed explicitly, the decision function will be denoted as $\delta^*(u_s)()$ or it is characterized by

$$\delta^*(u_s)(x)=v^* \rightarrow \max_v E_{u_e|(x,u_s)} G(u_e,P^n(v))= E_{u_e|(x,u_s)} G(u_e,P^n(v^*)).$$

The composition of δ and χ specifies the primary function of the institutional level management, transformation of an observed value of the external input into a global goal, operationalized one of the organizational goal, i.e.,

$$\chi\delta:X \rightarrow \hom(Y,Re).$$

The ideal primary function is, then, given by $\chi\delta^*$. Fig. 10.1 summarizes relations among the functions introduced above, where ϕ will be introduced below.

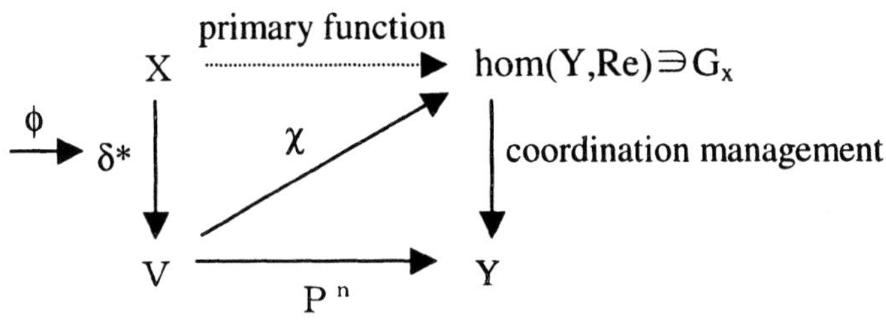

Figure10.1 Adaptation mechanism

It should be clear that if a valid coordination is achieved, χ, valid coordination and P^n constitute a commutative diagram. This indicates that the normative image is faithfully realized by the organizational level valid coordination activity.

Using Fig.10.1 let us formalize a management structure of the institutional level.

Let

$$Y_a \subset \{\delta | \delta : X \rightarrow V\} = \hom(X,V)$$

where Y_a will be called *internal adaptive activity* output and is specified as the range of a function ϕ below.

Since the adaptive activity of the institutional level is to modify its primary function following the change of the environmental structure, this chapter will use Y_a as the set of internal outcomes of the institutional level management for analysis of its adaptive behavior.

Let the *observation scheme* of U_s be:

$$\eta : U_s \rightarrow X_s$$

where X_s is the set of observation outcomes of U_s .

Then, it should be noticed that the observation of the external input consists of two levels: Observation of a value and observation of a structural change. The former observation is applied to δ to produce a global goal. An observed structural change will be used for modification of δ .

For technical reasons η is assumed to be an onto function. This assumption does not yield any practical limitation. Then, since there is a one-to-one correspondence from $U_s/\ker\eta$ to X_s, $U_s/\ker\eta$ will be used for X_s when the role of η is to be emphasized.

Notice that in general for a function $\xi : X \rightarrow Y$, $\ker\xi \subset X \times X$ is an equivalence relation defined by

$$(x,x') \in \ker\xi \leftrightarrow \xi(x)=\xi(x').$$

Let

$$M_a=\{d | d : X_s \rightarrow Y_a\} = \hom(X_s,Y_a).$$

M_a is the set of decision variables of institutional level management, which will be called *decision set*. Since $d(x_s)$ specifies a decision function from X to V for each observation $x_s \in X_s$, i.e.,

$$d(x_s) : X \rightarrow V,$$

d represents an *adaptation strategy* of the institutional level management.

It must be noticed that χd displays adaptive behaviors in two ways. If $u_s \in U_s$ is given, $d(\eta(u_s))(x)=v$ determines an adaptive behavior with respect to the change of u_e, and $d(\eta(u_s))=\delta$ displays an adaptive behavior with respect to the change of u_s. The former adaptation can be called tactical adaptation while the latter can be called strategic adaptation. This chapter will deal with the strategic adaptation. Notice, then, that the latter adaptation necessarily implies the former one in our formulation.

The process

$$P_a:M_a \times X_s \to Y_a$$

of the institutional level management can be expressed as:

$$P_a(d,x_s)=d(x_s).$$

Let

$$\phi:U_s \to Y_a$$

be a specification function of an optimal decision function such that for each $u_s \in U_s$

$$\phi(u_s)= \delta^*(u_s):X \to V$$

or

$$\phi(u_s)(x)=v^* \to \max_v E_{u_e|(x,u_s)} G(u_e,P^n(v))= E_{u_e|(x,u_s)} G(u_e,P^n(v^*)).$$

Furthermore, ϕ is assumed surjective or $Y_a \cong U_s/\ker\phi$.

Since ϕ always produces an optimal decision function in Y_a or a set point responding to the change of the external input structure, the goal (target) of the institutional level management is to realize it. Then,

goal of the institutional level management=realization of ϕ.

ϕ will be called *adaptation goal generator*. It should be noticed that ϕ is a composition of the organizational goal G, the normative model P^n and the intelligence Pr(); i.e., it presents a composite function of these components with respect to the adaptive activity of the institutional level. ϕ is the most important function of this chapter.

Conceptually, the management problem of the institutional level is to find $d \in M_a$ such that

$$|d(\eta(u_s)) - \phi(u_s)|$$

is minimized. This requirement will be formalized as *adaptation goal controllability*.

Finally, we have a goal seeking model of the institutional level as follows:

$$\text{goal seeking structure of institutional level} = \langle M_a, X, U_s, \underline{G}_x, P_a, G_a \rangle$$

where

$$\underline{G}_x = \{G_x | G_x : Y \to Re\},$$

and $G_a : M_a \times X \times U_s \times \underline{G}_x \to Re$ is

$$G_a(d, x, u_s, G_x) = \text{distance}(\chi[d\,(\eta(u_s))(x)], \chi[\phi(u_s)(x)])$$

and distance: $\underline{G}_x \times \underline{G}_x \to Re$ is a proper distance function.

Since

$$M_a = \hom(X_s, Y_a),$$

$$X_s \cong U_s / \ker\eta$$

and

$$Y_a \cong U_s / \ker\phi,$$

the essential structure to determine the institutional level activity is $\langle U_s, \phi, \eta \rangle$. For the sake of convenience, in this chapter the following form is used:

$$\text{adaptation structure} = \langle M_a, U_s, \phi, \eta \rangle.$$

Actually, subsequent arguments will be developed assuming $X_s = U_s/\ker\eta$ and $Y_a = U_s/\ker\phi$.

Notice that the adaptation structure is defined on function spaces, $M_a = \hom(X_s, Y_a)$, $Y_a \subset \hom(X, V)$, which shows clearly that the adaptive activity is a meta-managerial one.

The total structure of the institutional level management is shown in Fig. 10.2.

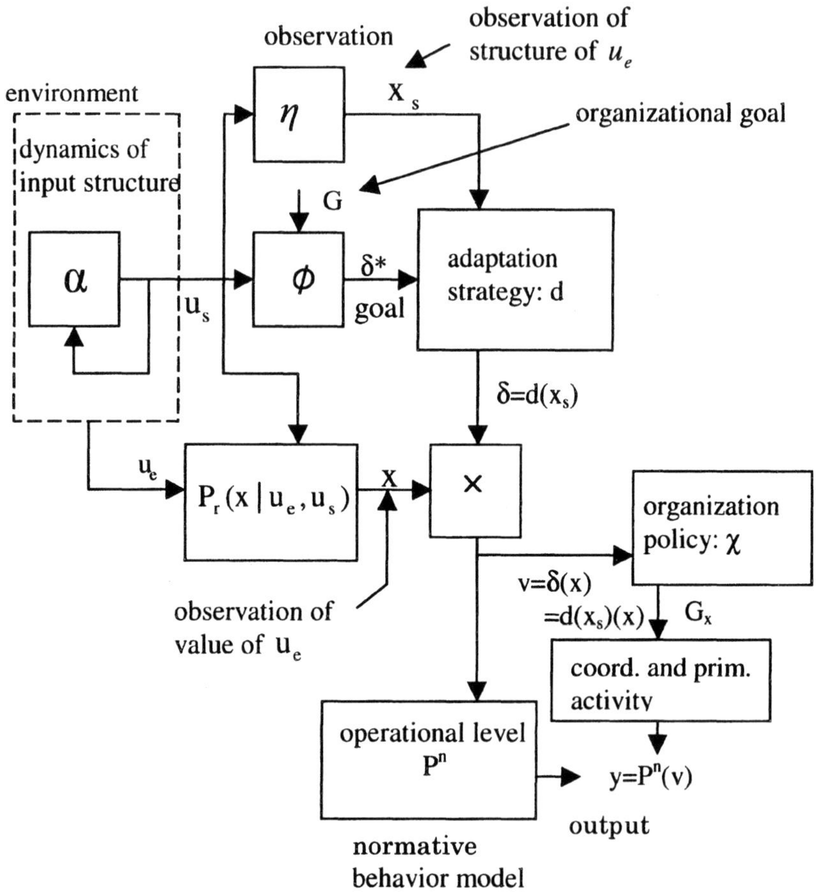

Figure 10.2 Institutional level management

c. Adaptation Goal Controllability and ϕ-α consistency

Using the above notations two basic concepts can be formalized for adaptation.

Definition 10.2 Adaptation goal controllability

An adaptation structure S_a=<M_a,U_s, ϕ, η> is said to have *adaptation goal controllability* property if and only if there exists an adaptation strategy d:X_s → Y_a such that ϕ is factorized by η and d, i.e.,

$$d\eta(u_s)= \phi(u_s)$$

holds for all $u_s \in U_s$. In particular, if S_a has the adaptation goal controllability property with respect to d, S_a is called an effective adaptation structure with respect to d.

The adaptation goal controllability property satisfies the following commutative diagram:

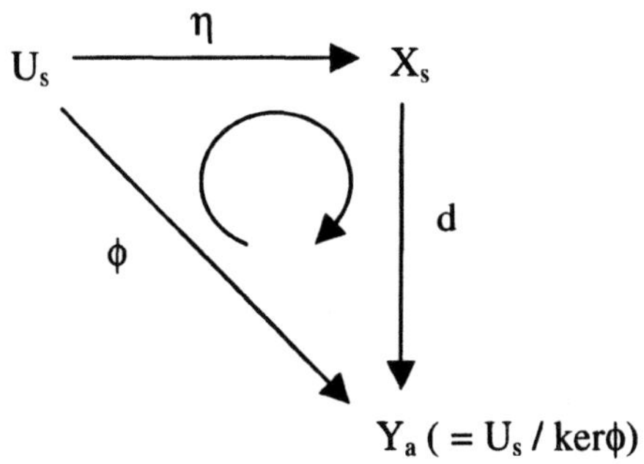

Figure 10.3 Adaptation goal controllability property

It should be noticed that since $\eta(u_s)=[u_s]_\eta$ and $\phi(u_s)=[u_s]_\phi$, d is uniquely determined as $d([u_s]_\eta)=[u_s]_\phi$, where $u_s \in U_s$ is arbitrary, if the adaptation goal controllability holds with respect to d. Then, this property is satisfied if $\eta(u_s)=\eta(u_s')$ implies $Pr(u_e|x,u_s)=Pr(u_e|x,u_s')$,

If the adaptation goal controllability holds, the following is true:

$$(\exists d)(\forall u_s)(\phi(u_s)=d\eta(u_s))$$

or $d\eta(u_s)(x)=\delta$ yields an optimal decision function. The above relation is a little bit too strong. What is required is the following relation:

$$(\forall u_s)\ (\exists d)(\phi(u_s)=d\eta(u_s)).$$

This chapter will develop a theory based on the strong condition.

As mentioned in the introduction of this chapter, this chapter will discuss a type of adaptation which maintains a desirable state of an organization confronted with a structural change of the environment. The desirable state (set point) of an organization of this chapter is that the organization keeps the adaptation goal controllability property, that is, ϕ is factorized by an element in M_a and the observation mechanism.

It is clear that the desired adaptation cannot be realized unless some consistency between the adaptation goal generator and the dynamics of the environment exists. The following ϕ-α consistency is introduced for that purpose.

Definition 10.3 φ-α consistency

Let the adaptation goal generator be $\phi: U_s \rightarrow Y_a$ and the dynamics of the structure of the external input be $\alpha: U_s \rightarrow U_s$. When φ and α satisfies the following relation

$$\ker\phi \subset \ker(\phi\alpha),$$

φ and α are called consistent. The consistency will be called *φ-α consistency.*

φ-α consistency requires that if the difference between u_s and u_s' can be ignored from the viewpoint of the adaptation goal generator (i.e., $\phi(u_s)=\phi(u_s'))$, they can be also ignored at the next stage (i.e., $\phi(\alpha(u_s))=\phi(\alpha(u_s')))$. That is, the dynamics given by α is moderate or consistent with respect to φ. This is a natural requirement to realize valid adaptation specified by φ.

10.2 Internal Model Principle

In this section we will show a strong result to showing that the "internal model principle", holds if the adaptation goal controllability and φ-α consistency hold. That is, when the two conditions are satisfied, the institutional level can have a precise image of the environmental change. Then, the system can display an ideal adaptation to the change of the external input structure.

Proposition 10.1

Suppose an adaptation structure $S_a=<M_a, U_s, \phi, \eta>$ and an input structure $U_s=<U_s, \alpha>$ are given. Suppose φ-α consistency holds. If S_a is effective, there is $\beta: Y_a \rightarrow Y_a$, where $Y_a = U_s/\ker\phi$, such that the following holds:

$$(\forall u_s \in U_s)(\beta\phi(u_s)=\phi\alpha(u_s))$$

and for some $d \in M_a$

$$(\forall u_s)(\forall t)(\phi\alpha^t(u_s)=\beta^t d\eta(u_s))$$

where $t \in \{0,1,2,..\}$.

The proof is given in Appendix.

If β exists as in Prop.10.1, β is called *internal model of the dynamics* α with respect to ϕ and d. β satisfies the commutative diagram of Fig. 10.4.

If the institutional level management structure has an internal model β with respect to ϕ and d, S_a can adapt to the change of the external input structure using only the internal image β, that is, $\beta^t\delta_0$ is an optimum decision function at any time t if $\delta_0=d\eta(u_s)$ holds. The adaptation by using β will be called *adaptation by the internal model principle.*

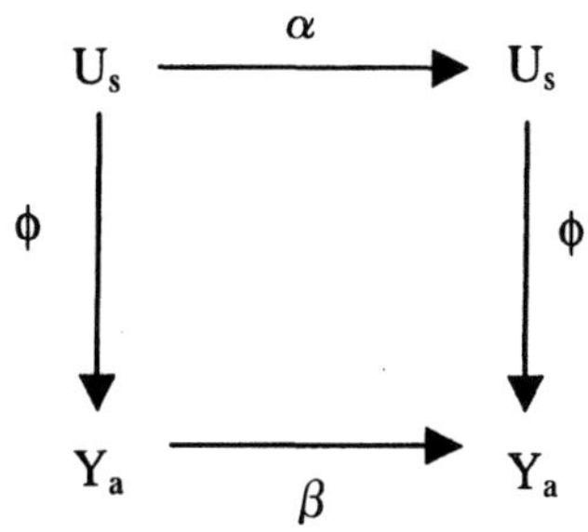

Figure 10.4 Internal model principle

Since an internal model is a faithful image of the dynamics of the external input, the following fact is intuitively clear.

Corollary 10.1
If an internal model exists, it is unique.

The proof is given in Appendix.

The converse of Prop. 10.1 also holds.

Proposition 10.2
Suppose an adaptation structure $S_a=<M_a, U_s, \phi, \eta>$ *has an internal model* $\beta:Y_a \to Y_a$ *of an input structure* $U_s=<U_s, \alpha>$ *with respect to an adaptation goal generator* ϕ *and an adaptation strategy d, i.e., adaptation by the internal model principle is valid where* $Y_a = U_s/\ker\phi$. *Then, the adaptation structure is effective and* ϕ*-*α *consistency holds.*

The proof is given in Appendix.

Then, Prop. 10.1 and Prop. 10.2 imply the following theorem.

Theorem 10.1

Suppose an adaptation structure S_a=<M_a, U_s, ϕ, η> and an input structure U_s=<U_s, α> are given. Then, ϕ-α consistency holds and S_a is effective if and only if adaptation by the internal model principle is valid.

Th. 10.1 shows that if ϕ-α consistency and the adaptation goal controllability are satisfied simultaneously, a strong result is induced.

10.3 Environmental Variety and Internal Variety

The environmental variety and the internal variety will be defined in terms of ϕ and α. The pair (ϕ,α) has a special meaning in investigation of variety. ϕ generates a goal (a set point) of the institutional management while α specifies the dynamics or changeability of the environmental structure. The varieties must be measured relative to (ϕ,α). Only the cardinality of U_s or Y_a cannot fully reflect the significance of variety.

The following simple lemma is of conceptual importance.

Lemma 10.1

Suppose an adaptation structure S_a=<M_a, U_s, ϕ, η> and an input structure U_s=<U_s, α> are given. Let $E(\phi,\alpha) \subset U_s \times U_s$ be

$$(\forall u_s, u_s' \in U_s)((u_s,u_s') \in E(\phi,\alpha) \leftrightarrow (\forall t)(\phi\alpha^t(u_s) = \phi\alpha^t(u_s'))).$$

Then, $E(\phi,\alpha)$ is an equivalence relation.

$E(\phi,\alpha)$ specifies an equivalence relation which says when the difference between two structure values u_s and u_s' can be ignored in the view of the adaptation goal generator ϕ.

Then, using $E(\phi,\alpha)$ an environmental variety can be introduced.

Definition 10.4 Environmental variety

$U_s/E(\phi,\alpha)$ is called *environmental variety* of an input structure U_s=<U_s, α> with respect to an adaptation structure S_a=<M_a, U_s, ϕ, η>.

Significance of the environmental variety is measured with respect to the goal generator ϕ of the institutional level.

An internal variety is naturally defined by M_a which is the set of alternatives to cope with change of the environmental structure.

Definition 10.5 Internal variety

Let an adaptation structure S_a=<M_a, U_s, ϕ, η> be given. Then, $M_a=\hom(U_s/\ker\eta, U_s/\ker\phi)$ is called *internal variety* of S_a.

The law of the requisite variety holds when the internal variety is not redundant. Since M_a depends on the observation scheme η, efficiency of M_a can be specified in terms of η. Non-redundancy of M_a in this sense is defined by the next S_a-optimality concept.

Definition 10.6 S_a-optimality

Let an adaptation structure be S_a=<M_a, U_s, ϕ, η>. Let an adaptation strategy be $d^* \in M_a$. Then, (d^*,η) is called S_a*-optimal* or *simply optimal* if and only if

(i) S_a satisfies the adaptation goal controllability property with respect to d^*, i.e., $\phi=d^*\eta$.

(ii) For every adaptation structure S_a'=<M_a', U_s, ϕ, η'> which satisfies the adaptation goal controllability property with respect to an adaptation strategy d' there is a natural mapping $\rho:U_s/\ker\eta' \rightarrow U_s/\ker\eta$ such that Fig. 10.5 is commutative.

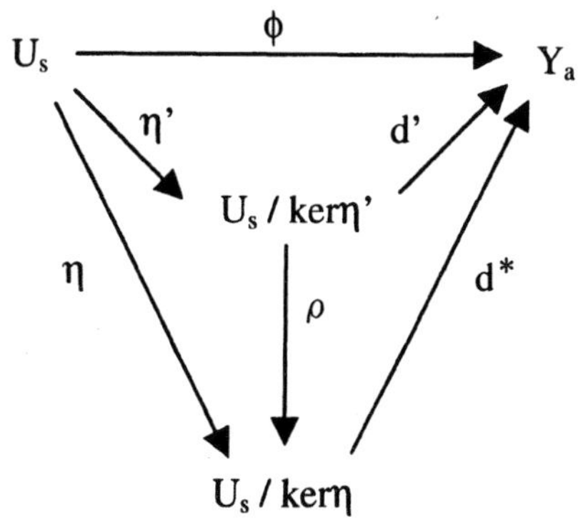

Figure 10.5 S_a-optimality

The condition (i) of Def. 10.6 requires that S_a is effective while the condition (ii) requires that the internal variety is efficient as Cor. 10.2 will imply.

Definition 10.7 efficient and effective Internal Variety

Suppose an adaptation structure S_a=<M_a, U_s, ϕ, η> is given. Then, the decision set M_a is called *efficient and effective internal variety* if and only if there is an adaptation strategy d^* in M_a such that (d^*,η) is S_a-optimal.

Corollary 10.2

Let an adaptation structure S_a=<M_a, U_s, ϕ, η> *be given. Suppose the decision set* M_a *is an efficient and effective internal variety. Then, if an adaptation structure* $S_a' = <M_a', U_s, \phi, \eta'>$ *is effective, there is a natural surjective map* $\xi : M_a' \rightarrow M_a$.

The proof is given in Appendix.

Cor. 10.2 indicates that the decision set M_a is more efficient than the decision set M_a' in the sense

$$|M_a| \leqq |M_a'|.$$

The following is a characterization of an efficient and effective internal variety.

Proposition 10.3

Suppose an adaptation structure $S_a = <M_a, U_s, \phi, \eta>$ *and an input structure* $U_s = <U_s, \alpha>$ *are given. Suppose* ϕ-α *consistency holds. Then, the decision set* M_a *is an efficient and effective internal variety if and only if* $ker\eta = E(\phi, \alpha)$.

The proof is given in Appendix.

The next lemma assures the uniqueness of an efficient and effective internal variety when U_s and ϕ are fixed.

Lemma 10.2

Suppose an adaptation structure $S_a = <M_a, U_s, \phi, \eta>$ *and an input structure* $U_s = <U_s, \alpha>$ *are given. Suppose* ϕ-α *consistency holds. Suppose the decision set* M_a *is an efficient and effective internal variety with respect to an adaptation strategy* $d^* \in M_a$, *i.e.,* $\phi = d^*\eta$. *Then*

(i) For any observation function $\eta' : U_s \rightarrow U_s/ker\eta'$, $hom(U_s/ker\eta', U_s/ker\phi)$ *is not an efficient and effective internal variety of* S_a *if* $ker\eta \neq ker\eta'$.

(ii) If (d', η) *is* S_a*-optimal where* $d' \in M_a$, *we have* $d^* = d'$, *i.e., the optimal adaptation strategy is unique.*

The proof is given in Appendix.

10.4 Variety Matching

Let us define orders on the varieties to examine the change of them in a quantitative way.

Definition 10.8 Order of environmental variety

The external variety $U_s/E(\phi,\alpha)$ of an input structure $U_s=<U_s, \alpha>$ with respect to an adaptation structure $S_a=<M_a, U_s, \phi, \eta>$ is said to be larger than the variety $U_s'/E(\phi', \alpha')$ of an input structure $U_s'=<U_s', \alpha'>$ with respect to an adaptation structure $S_a'=<M_a', U_s', \phi', \eta'>$ if the following holds:

(i) $U_s' \subset U_s$.

(ii) $(\forall\ u_s \in U_s-U_s')([u_s] \subset U_s-U_s')$ and $(\forall\ u_s \in U_s')([u_s] \subset U_s')$ where $[u_s] \in U_s/E(\phi,\alpha)$.

(iii) $E(\phi,\alpha) \cap (U_s' \times U_s') \subset E(\phi', \alpha')$.

If the conditions hold, we will write

$U_s/E(\phi,\alpha) \geqq U_s'/E(\phi', \alpha')$.

The relation $\geqq$ of Def. 10.8 is a partial order as shown by Prop. 10.4. The order is defined between two external varieties whose structures are similar or the larger variety is an extension of the smaller one. The conditions (i) and (iii) are natural for any extension. The condition (ii) requires that the extension is done without yielding any conflict between the original structure and the extended one.

The following is a technical fact.

Lemma 10.3

For every $u_s \in U_s'$, if $U_s/E(\phi,\alpha) \geqq U_s'/E(\phi', \alpha')$ holds, we have $[u_s] \subset [u_s]'$.

The proof is given in Appendix.

The following proposition asserts that $\geqq$ of the external variety specifies an order relation for the variety.

Proposition 10.4

$\geqq$ *defined in* Def. 10.8 *is a partial order.*

The proof is given in Appendix.

The following lemma shows a condition when an order $\geqq$ can hold.

Lemma 10.4

Suppose an adaptation structure S_a=<M_a, U_s, ϕ, η>, *input structures* U_s=<U_s, α> *and* U_s'=<U_s, α'> *are given. If* $(\forall t)(\exists t')(\forall u_s \in U_s)(\alpha^{t'}(u_s)=\alpha'^{t}(u_s))$ *holds, then*

$$U_s/E(\phi,\alpha) \geqq U_s'/E(\phi,\alpha')$$

holds.

The proof is given in Appendix.

Since in Lemma 10.4 the behavior of U_s is an extension of that of U_s' along the time coordinate, it is naturally expected that the relation of Lemma 10.4 should hold between them.

An order relation for the internal variety is specified as below..

Definition 10.9 Order of internal variety

Suppose adaptation structures S_a=<M_a, U_s, ϕ, η> and S_a'=<M_a', U_s', ϕ', η'> are given.

The internal variety M_a is said larger than the internal variety M_a' if

(i) $U_s' \subset U_s$

(ii) $|M_a| \geqq |M_a'|$.

If these conditions hold, we will write

$$M_a \geqq M_a'.$$

$\geqq$ of the internal variety is a pre-order. It is clear that this order is also defined assuming extensions of internal varieties.

When the size of the underlying set U_s is uncontrollable, the institutional level management has at least two ways to increase $|M_a|=|\hom(U_s/\ker\eta, U_s/\ker\phi)|$: The first is to refine the equivalence relation $\ker\eta$ or to extend the size of $U_s/\ker\eta$; the second is to extend the size of $U_s/\ker\phi(=Y_a)$. The following proposition shows the variety increase of the former method. It should be noticed that the observation scheme η can be a "controllable" parameter of the institutional level while Y_a may be not.

Proposition 10.5

Suppose adaptation structures S_a=<M_a, U_s, ϕ, η> *and* S_a'=<M_a', U_s', ϕ', η'> *are given. Suppose* $U_s=U_s'$ *and* $U_s/\ker\phi= U_s'/\ker\phi'$. *Then, if* $\ker\eta \subset \ker\eta'$ *holds, we have*

$$M_a \geqq M_a'.$$

The proof is given in Appendix.

Finally, we have a formal statement of the law of requisite variety.

Theorem 10.2

Suppose adaptation structures S_a=<M_a, U_s, ϕ, η> *and* S_a'=<M_a', U_s', ϕ', η'>, *and input structures* U_s=<U_s, α> *and* U_s'=<U_s', α'> *are given. Suppose* ϕ-α *consistency and* ϕ'-α' *consistency hold.*

Suppose $U_s' \subset U_s$ *and* $\phi'=\phi|U_s'$. *If* M_a *and* M_a' *are efficient and effective internal varieties of* S_a *and* S_a', $U_s/E(\phi,\alpha) \geqq U_s'/E(\phi', \alpha')$ *implies* $M_a \geqq M_a'$.

The proof is given in Appendix.

The above results interprets the law of requisite variety in our formal framework. Suppose the internal variety is efficient and effective in the sense of Def. 10.7. Suppose the environmental variety increases from $U_s'/E(\phi', \alpha')$ to $U_s/E(\phi,\alpha)$. Then in order to keep the management effective (S_a-optimal is satisfied), the institutional level management has to increase the internal variety from home($U_s'/\ker\eta'$, $U_s'/\ker\phi'$) to home($U_s/\ker\eta$, $U_s/\ker\phi$). Of course this statement does not holds if the internal variety is redundant.

APPENDIX OF CHAPTER 10

Proof of Proposition 10.1

We will first construct $\beta : Y_a \rightarrow Y_a$ and then show that β satisfies the conditions.

Let $\delta \in Y_a$ be arbitrary. If there is $u_s \in U_s$ such that $\phi(u_s)=\delta$, we define β by

$$\beta(\delta)=\phi\alpha(u_s).$$

If there is no such u_s in U_s then let $\beta(\delta)$ be arbitrary in Y_a. Then, β is well-defined: Indeed, let u_s and u_s' be in U_s such that $\phi(u_s)=\delta=\phi(u_s')$ or $(u_s,u_s') \in \ker\phi$, then since ϕ and α are consistent, we have $\phi\alpha(u_s) = \phi\alpha(u_s')$ and hence β is well-defined. Because of the construction way of β

$$\beta\phi(u_s)=\phi\alpha(u_s)$$

for any $u_s \in U_s$.

Next, we prove β satisfies the second condition by induction on t. Due to the adaptation goal controllability property there exists $d \in M_a$ such that $\phi(u_s)=d\eta(u_s)$ holds for every $u_s \in U_s$. Furthermore, since β satisfies

$$\phi\alpha(u_s) = \beta\phi(u_s) = \beta d\eta(u_s),$$

the cases of t=0 and 1 obviously hold.

Now assume the statement holds for t=i−1. Then we have

$$\beta^i\phi(u_s) = \beta(\beta^{i-1}\phi(u_s)) = \beta\phi\alpha^{i-1}(u_s).$$

By the first property of β it follows that

$$\beta\phi\alpha^{i-1}(u_s) = \phi\alpha\alpha^{i-1}(u_s) = \phi\alpha^i(u_s)$$

and the proof is complete.

Q.E.D.

Proof of Corollary 10.1

Suppose there are β and β' such that $\beta\phi(u_s)=\phi\alpha(u_s)$ and $\beta'\phi(u_s)=\phi\alpha(u_s)$ for every $u_s \in U_s$. Then $\beta\phi(u_s)=\beta'\phi(u_s)$. Since ϕ is surjective, it follows that $\beta=\beta'$.

Q.E.D.

Proof of Proposition 10.2

Suppose the assumptions are satisfied and β is an internal model of U_s with respect to ϕ and d. From the assumption we have

$$(\forall\ t \in \{0, 1,2... \})(\beta^t d\eta(u_s)= \phi\alpha^t(u_s))$$

for every $u_s \in U_s$. By setting t=0 we have $d\eta(u_s)= \phi(u_s)$ holds and hence S_a has the adaptation goal controllability property.

Now we have only to show $\ker\phi \subset \ker\phi\alpha$. Let $(u_s,u_s') \in \ker\phi$. Then we have

$$\phi\alpha(u_s)= \beta\,\phi(u_s) = \beta\,\phi(u_s')$$

$$= \phi\alpha(u_s')$$

since $\beta\,\phi(u_s)= \phi\alpha(u_s)$ holds for every $u_s \in U_s$ due to the property of β. Consequently, we have $\ker\phi \subset \ker\phi\alpha$.

Q.E.D.

Proof of Corollary 10.2

Let $d \in \hom(U_s/\ker\eta, Y_a)$ be arbitrary. Then, using ρ of Fig. 10.5, $d' \in \hom(U_s/\ker\eta', Y_a)$ is defined as:

$$d'(x_s)=d(\rho x_s)$$

for any $x_s \in U_s/\ker\eta'=X_s'$. The correspondence ξ is given by $\xi(d')=d$.

Q.E.D.

Proof of Proposition 10.3

Let us consider the if part first. Let $\ker\eta = E(\phi,\alpha)$. We will show that there is a an adaptation strategy d^* in $\hom(U_s/E(\phi,\alpha), Y_a)$ such that (d^*,η) is S_a-optimal.

Let us define $d^*:U_s/E(\phi,\alpha) \rightarrow Y_a$ by

$$d^*([u_s])= \phi(u_s)$$

for every $[u_s] \in U_s/E(\phi,\alpha)$.

Then (d^*,η) is an S_a-optimal. Indeed, let $(u_s, u_s') \in E(\phi,\alpha)$, then by the definition α

$$(\forall\ t \in \{0, 1, 2,... \})(\phi\alpha^t(u_s) =\phi\alpha^t(u_s'))$$

holds, which implies $\phi(u_s)= \phi(u_s')$. Hence, d^* is well-defined. Furthermore, $S_a=<M_a, U_s, \phi, \eta>$ clearly has the adaptation goal controllability property with respect to d^* since

$$d^*\eta(u_s)=d^*([u_s])= \phi(u_s)$$

holds for every $u_s \in U_s$.

Suppose (d', η') is such that $d'\eta'(u_s)= \phi(u_s)$ holds for every $u_s\in U_s$, where $\eta': U_s \to U_s/\ker\eta'$ and $d':U_s/\ker\eta' \to Y_a$. We will show that there is a natural map $\rho:U_s/\ker\eta' \to U_s/\ker\eta$ such that Fig. 10.5 is commutative.

Let $(u_s,u_s')\in \ker\eta'$. Since $S_a'=<M_a, U_s, \phi, \eta'>$ has the adaptation goal controllability property with respect to d', it follows from Prop. 10.1 that there is $\beta :Y_a \to Y_a$ such that

$$(\forall\ t \in \{0, 1, 2,\dots\})(\beta^t d'\eta'(u_s)= \phi\alpha^t(u_s)).$$

On the other hand, $\eta'(u_s)= \eta'(u_s')$ implies

$$(\forall\ t \in \{0, 1, 2\dots\})(\beta^t d'\eta'(u_s) = \beta^t d'\eta'(u_s')).$$

Thus,

$$\phi\alpha^t(u_s)= \beta^t d'\eta'(u_s)= \beta^t d'\eta'(u_s')= \phi\alpha^t(u_s')$$

holds for every t and hence $(u_s,u_s') \in E(\phi,\alpha)=\ker\eta$. Consequently, we have $\ker\eta' \subset \ker\eta$ and hence we can define a natural map $\rho:U_s/\ker\eta' \to U_s/\ker\eta$ by

$$\rho([u_s]')=[u_s]$$

where $[u_s]$ and $[u_s]'$ denote the equivalence classes of u_s with respect to $\ker\eta$ and $\ker\eta'$, respectively.

Finally we will show ρ makes Fig. 10.5 commutative. Since

$$(\forall\ u_s \in U_s)(\rho\eta'(u_s)= \rho([u_s]')=[u_s]= \eta(u_s))$$

holds and since S_a and S_a' have the adaptation goal controllability property,

$$(\forall\ u_s \in U_s)(d\rho\eta'(u_s) =d\eta(u_s)= \phi(u_s) =d'\eta'(u_s))$$

holds. Hence, we have $d\rho\eta'=d'\eta'$. Since η' is surjective, we have $d\rho=d'$.

Next, we will prove the only if part. Let M_a be an efficient and effective internal variety. By the definition of the efficiency and effectiveness there is an adaptation strategy $d^*:U_s/\ker\eta \rightarrow Y_a$ such that (d^*,η) is an S_a -optimal. We will show $\ker\eta=E(\phi,\alpha)$.

Suppose $(u_s,u_s')\in\ker\eta$ holds. Since S_a has the adaptation goal controllability property with respect to d^*, by Prop. 10.1 there is $\beta:Y_a \rightarrow Y_a$ such that

$$(\forall\ t\in\{0,1,2,\dots\})(\beta^t d^*\eta(u_s)=\phi\alpha^t(u_s))$$

holds for every $u_s\in U_s$. Consequently, we have

$$(u_s,u_s')\in\ker\eta$$

$$\rightarrow\eta(u_s)=\eta(u_s')$$

$$\rightarrow(\forall\ t\in\{0,1,2,\dots\})(\beta^t d^*\eta(u_s)=\beta^t d^*\eta(u_s'))$$

$$\rightarrow(\forall\ t\in\{0,1,2,\dots\})(\phi\alpha^t(u_s)=\phi\alpha^t(u_s'))$$

$$\rightarrow(u_s,u_s')\in E(\phi,\alpha).$$

On the other hand, suppose that $(u_s,u_s')\in E(\phi,\alpha)$ holds. Then, the definition of $E(\phi,\alpha)$ implies

$$(\forall\ t\in\{0,1,2,\dots\})(\phi\alpha^t(u_s)=\phi\alpha^t(u_s')).$$

By setting t=0 we have $\phi(u_s)=\phi(u_s')$. Now since

$$(\forall\ u_s\in U_s)(1\phi(u_s)=\phi(u_s))$$

trivially holds, where 1 denotes an identity on Y_a, $\langle M_a,U_s,\phi,\eta'\rangle$ has the adaptation goal controllability property with respect to 1 where $\eta'=\phi$. Since (d^*,η) is S_a-optimal, by the condition (ii) of Def. 10.6 there is ρ: $\phi(U_s)\rightarrow U_s/\ker\eta$, where $U_s/\ker\eta'=U_s/\ker\phi\cong\phi(U_s)=Y_a$, such that

$$(\forall\ u_s\in U_s)(\rho\phi(u_s)=\eta(u_s))$$

holds. Consequently, it follows from $\phi(u_s)= \phi(u_s')$ that

$$\eta(u_s) = \rho\phi(u_s) = \rho\phi(u_s') = \eta(u_s'),$$

which implies $(u_s, u_s') \in \ker\eta$.

Q.E.D.

Proof of Lemma 10.2

(i) Obvious from Prop. 10.3.

(ii) Let d and d' $\in$ hom($U_s/\ker\eta, Y_a$) where (d, η) and (d', η) are S_a-optimal with respect to (ϕ,α). Then, S_a has the adaptation goal controllability property with respect to d and d'. Prop. 10.1 implies that there are $\beta : Y_a \to Y_a$ and $\beta' : Y_a \to Y_a$ such that

$$(\forall\ u_s \in U_s)(\forall\ t \in \{(0,1,2\ldots)\}(\beta^t d\eta(u_s) = \phi\alpha^t(u_s))$$

and

$$(\forall\ u_s \in U_s)(\forall\ t \in \{0,1,2\ldots\})(\beta'^t d'\eta(u_s) = \phi\alpha^t(u_s)).$$

By setting t=0 it follows that $d\eta=\phi=d'\eta$. Since the domain of d and d' is $\eta(U_s)$, it implies d=d'.

Q.E.D.

Proof of Lemma 10.3

Let $u_s \in U_s'$, and $v \in [u_s]$ be arbitrary. Since $u_s \in U_s'$ implies $[u_s] \subset U_s'$ by the condition (ii) of Def. 10.8, we have $v \in U_s'$. Thus $(u_s,v) \in E(\phi,\alpha) \cap (U_s' \times U_s')$ holds and hence it follows from the condition (iii) of Def. 10.8 that $(u_s, v) \in E(\phi', \alpha')$. Consequently, we have $v \in [u_s]'$.

Q.E.D.

Proof of Proposition 10.4

i) $\leqq$ will be shown to be reflexive.

$U_s/E(\phi,\alpha) \leqq U_s/E(\phi,\alpha)$ holds since the conditions (i) and (iii) obviously hold while (ii) holds as a tautology if $U_s - U_s'$ is empty.

ii) $\leqq$ will be shown to be anti-symmetric.

If $U_s/E(\phi,\alpha) \leqq U_s'/E(\phi', \alpha')$ and $U_s'/E(\phi', \alpha') \leqq U_s/E(\phi,\alpha)$ hold, then the conditions (i) and (iii) imply $U_s'/E(\phi', \alpha')=U_s/E(\phi,\alpha)$.

iii) $\leqq$ is transitive:

Suppose $U_s/E(\phi,\alpha) \geqq U_s'/E(\phi', \alpha')$ and $U_s'/E(\phi', \alpha') \geqq U_s''/E(\phi'',\alpha'')$ hold. Then we have

$$(\forall u_s \in U_s - U_s')([u_s] \subset U_s - U_s') \text{ and } (\forall u_s \in U_s')([u_s] \subset U_s') \tag{A}$$

and

$$(\forall u_s \in U_s' - U_s'')([u_s]' \subset U_s' - U_s'') \text{ and } (\forall u_s \in U_s'')([u_s]' \subset U_s''), \tag{B}$$

where $[u_s]$ and $[u_s]'$ denote the equivalence classes of u_s with regard to $E(\phi,\alpha)$ and $E(\phi', \alpha')$, respectively.

It is only required to verify

$$(\forall u_s \in U_s - U_s'')([u_s] \subset U_s - U_s'')$$

and

$$(\forall u_s \in U_s'')([u_s] \subset U_s'')$$

in order to show

$$E(\phi,\alpha) \leqq E(\phi'',\alpha'').$$

Let $u_s \in U_s - U_s''$ be arbitrary. Then we have either $u_s \in U_s - U_s'$ or $u_s \in U_s' - U_s''$. If $u_s \in U_s - U_s'$ holds, then (A) implies $[u_s] \subset U_s - U_s'$. Since it follows from $U_s'' \subset U_s'$ that $U_s - U_s' \subset U_s - U_s''$, we have $[u_s] \subset U_s - U_s''$. Similarly, if $u_s \in U_s' - U_s''$ holds then (B) implies $[u_s]' \subset U_s' - U_s''$. Since $u_s \in U_s'$ holds, Lemma 10.3 implies $[u_s] \subset [u_s]'$. It follows that $[u_s] \subset U_s' - U_s''$, which yields $[u_s] \subset U_s - U_s''$, since $U_s' - U_s'' \subset U_s - U_s''$ holds. Consequently, we have

$$(\forall u_s \in U_s - U_s'')([u_s] \subset U_s - U_s'').$$

Let $u_s \in U_s''$ be arbitrary. Then we have $[u_s]' \subset U_s''$ by (B). By applying Lemma 10.3 again, we have $[u_s] \subset U_s''$.

Q.E.D.

Proof of Lemma 10.4

Since the conditions (i) and (ii) of Def. 10.8 clearly hold, it is sufficient to show $E(\phi,\alpha) \subset E(\phi,\alpha')$.

Let $(u_s, u_s') \in E(\phi,\alpha)$. Then, by definition we have

$$(\forall t \in \{0, 1, 2 \dots\})(\phi\alpha^t(u_s) = \phi\alpha^t(u_s')).$$

We will show

$(\forall t \in \{0, 1, 2\ldots\})(\phi\alpha''^t(u_s) = \phi\alpha''^t(u_s'))$.

Since for every $t \in \{0,1,2\ldots\}$ there is $t' \in \{0,1,2\ldots\}$ such that for every $u_s \in U_s$ $\alpha''^t(u_s) = \alpha^{t'}(u_s)$ holds, we have

$$\phi\alpha''^t(u_s) = \phi\alpha^{t'}(u_s) = \phi\alpha^{t'}(u_s') = \phi\alpha''^t(u_s')).$$

Q.E.D.

Proof of Proposition 10.5

We will show that $|\hom(U_s/\ker\eta, Y_a)| \geqq |\hom(U_s/\ker\eta', Y_a)|$ by verifying that there is an injection from $\hom(U_s/\ker\eta', Y_a)$ to $\hom(U_s/\ker\eta, Y_a)$. Since $\ker\eta \subset \ker\eta'$ holds, there is a natural map $\rho: U_s/\ker\eta \rightarrow U_s/\ker\eta'$. Now let us define $F: \hom(U_s/\ker\eta', Y_a) \rightarrow \hom(U_s/\ker\eta, Y_a)$ by $F(d') = d'\rho$ for every $d' \in \hom(U_s/\ker\eta', Y_a)$. Then F is injective. Indeed, if $F(d') = F(d'')$ holds, then we have $d'\rho = d''\rho$, which implies $d' = d''$ since ρ is surjective.

Q.E.D.

Proof of Theorem 10.2

Let us assume $U_s/E(\phi,\alpha) \geqq U_s'/E(\phi', \alpha')$. By the assumption and Prop. 10.3 we have $\ker\eta = E(\phi,\alpha)$ and $\ker\eta' = E(\phi', \alpha')$. We are only required to show that $|\hom(U_s/\ker\eta, Y_a)| \geqq |\hom(U_s'/\ker\eta', Y_a')|$ since $U_s' \subset U_s$ holds by the assumption.

We will verify it by constructing an injection

$$F: \hom(U_s'/\ker\eta', Y_a') \rightarrow \hom(U_s/\ker\eta, Y_a).$$

Since ϕ and ϕ' are surjective, $\phi(U_s) = Y_a$ and $\phi'(U_s') = Y_a'$ hold. $U_s' \subset U_s$ implies $Y_a' \subset Y_a$ and hence there is an inclusion $i: Y_a' \rightarrow Y_a$.

Now let us define $\rho: U_s/\ker\eta \rightarrow U_s'/\ker\eta'$ by

$$\rho([u_s]) = \begin{cases} [u_s]' \text{ if } u_s \in U_s' \\ [u_s^*]' \text{ otherwise} \end{cases}$$

where u_s^* is arbitrary in U_s' while $[u_s]$ and $[u_s]'$ denote the equivalence classes of u_s with respect to $\ker\eta = E(\phi,\alpha)$ and $\ker\eta' = E(\phi', \alpha')$, respectively. Then, the definition is valid. Indeed, let u_s and v_s be in U_s such that $[u_s] = [v_s]$. We can verify $\rho([u_s]) = \rho([v_s])$ as below.

Suppose $u_s \in U_s - U_s'$. Then, by definition we have $[u_s] \subset U_s - U_s'$, which implies $[v_s] \subset U_s - U_s'$ and hence $v_s \in U_s - U_s'$.

Similarly if $u_s \in U_s'$ holds, then $v_s \in U_s'$. Consequently, we have either u_s, $v_s \in U_s - U_s'$ or u_s, $v_s \in U_s'$ when $[u_s]=[v_s]$.

If u_s and v_s are in $U_s - U_s'$, then we have

$$\rho([u_s]) = [u_s^*]' = \rho([v_s]).$$

On the other hand, if u_s and v_s are in U_s', then we have $(u_s,v_s) \in E(\phi', \alpha')$. That is, $[u_s]'=[v_s]'$ holds, because $[u_s] = [v_s]$ and $E(\phi,\alpha) \cap (U_s' \times U_s') \subset E(\phi', \alpha')$(condition (iii) of Def. 10.8).
Hence, it follows

$$\rho([u_s])=[u_s]'=[v_s]'=\rho([v_s]).$$

Furthermore, ρ is surjective: Let $[u_s]' \in U_s'/\ker\eta'$ be arbitrary. Then there is $[u_s] \in U_s/\ker\eta$ and we have $\rho([u_s])=[u_s]'$.

Now let us define F by

$$F(d') = id'\rho$$

for every $d' \in$ home$(U_s'/\ker\eta', Y_a')$ where $i: Y_a' \to Y_a$ is the embedding mapping. Then F is clearly a function from $\hom(U_s'/\ker\eta', Y_a')$ to $\hom(U_s/\ker\eta, Y_a)$. And F is injective: Suppose $F(d')=F(d'')$. Then by definition we have $id'\rho=id''\rho$. Since i is injective and ρ is surjective we have $d'=d''$.

Q.E.D.

CHAPTER 11 COMPUTATIONAL APPROACH TO ORGANIZATION

GARBAGE CAN MODEL

In previous chapters we have investigated an organization (or an organization model) in qualitative organization theory using non-numerical mathematics. There is another way to study an organization in a formal way. It is called the computational approach. The approach builds a computer model of an organization and explores organizational properties examining the model's computer behavior. Since a computer model is a formal representation of an object, the approach is conceptually akin to our approach. Both approaches try to develop a solid theory for an organization using operational languages like computer programs or non-numerical mathematics.

The qualitative mathematics approach and the computational approach are complementary. The former approach provides a grand theory in formal terms while the latter approach can examine detailed behaviors of a given organization in concrete terms. The former is more concerned with the structure of an organization while the latter is more concerned with the process behavior of an organization. A qualitative mathematics theory presents a general view while a computational theory presents a concrete view. In this sense both approaches are necessary for a formal theory of an organization.

Ref. [18] categorizes organization into four models: Rational, bureaucratic, political and "garbage can".

The organizations discussed in the previous chapters are certainly of a rational category.

According to Ref. [18] the bureaucratic model is characterized by standard operating procedures. Chapter 9 assumed the bureaucratic model for the operational level. The standard operating procedures are programmed ones. The primary purpose of the organization in this category is to survive;

uncertainty reduction is the principal goal. Policies are chosen that are incrementally different from the past.

Ref. [18] also says that the political category assumes that organizational decisions result from political competitions; key players are involved in a game of influence, bargaining and power. Organizational outcomes are determined by the beliefs and goals of players, their skills in playing games, the resources they bring to bear and the limits on their attention and power.

The garbage can category is the most anarchical one which is at the opposite end of the rational category [10]. Ref. [10] says that "although it may be convenient to imagine that choice opportunities lead first to the generation of decision alternatives, then to an examination of the consequences, then to an evaluation of these consequences in terms of objectives, and finally, to a decision, this type of model is often a poor description of what actually happens". The garbage can theory insists that "a decision is an outcome or interpretation of several relatively independent streams within an organization." The streams are concerned with problems, choice opportunities, participants and solutions.

Since the garbage can model presents an antithesis of the development that the preceding chapters have been engaged in, it is quite instructive to compare the developed theory to the garbage can theory; or to examine how positively or negatively the rational model of Fig. 1.4 can be explained in terms of the garbage can model and vice versa.

There is another important point about the garbage can model. That is, in Ref. [10] it is formalized as a computational model and is investigated as such. It provides a typical example of the computational approach. This section will interpret the garbage can model in the rational model of this book, formalize it in our approach or formalize it in the set theory using systems concepts and present a computational implementation of the formulation in Prolog. As a matter of fact a set theoretic formulation and its Prolog implementation is the computational approach of our formalism. Prolog is used because it can represent a model in the set theory in a straightforward way. This topic will be illustrated but not explored in this book [33].

11.1 Computational Approach to Organization Theory

Ref. [1] summarizes a model of the COT (the computational organization theory) as an object which is a collections of process and intelligent adaptive agents that are task oriented, socially situated, technologically bound and continuously changing. It also says that an organization is modeled by three components: Model of agents, model of tasks, and model of internal structure. Then, an organization is characterized by agents, resources/tasks, and skills.

This image is certainly not incompatible with the model of the OCT (organizational cybernetic theory) we have so far discussed.

An agent of the COT certainly corresponds to a goal seeking player of the OCT. In the OCT agents are categorized into three classes: An agent on the operational level, an agent on the organizational level and an agent on the institutional level. In the OCT an agent is a manager or a decision maker whose behavior is explained by a goal seeking model which consists of a goal seeking problem and a decision principle.

The task of the COT corresponds to a goal seeking activity in the OCT where it is particularly named after its functions as optimization, coordination and adaptation according to their hierarchical position.

The skill of the COT corresponds to the process model of the OCT. In particular, $\underline{\tau}$, process time of Chapter 9 directly represents skill.

The internal structure of the OCT is a hierarchical one consisting of three layers and the functions of each layer are explicitly defined.

In addition to the facts that a model of an organization is more structured in the OCT than in the COT and since the OCT is described in the set theory terms, formal theories of the OCT can be deduced, there are two things which distinguish the COT from the OCT.

First, the COT emphasizes a dynamics of an agent or of an organization. As mentioned in the previous chapters, since variables of the OCT which describe an organization or a goal seeking model (or an agent) can be time functions, a model of the OCT can naturally cover a dynamics of an organization in the sense of the functional analysis. It is clear, however, that details of a dynamics cannot be treated fully in the functional analysis. Sometimes we need a representation by a differential (difference) equation just for the sake of illustration. The COT has an advantage over the OCT in investigating the dynamics in detail.

Second, although constraints of management resources are critical problems to management of an organization, they are implicitly handled as constraints to a goal seeking activity in the OCT. As Sec. 11.2 shows, the problem can be easily treated in the COT.

According to Ref. [1] methodologies adopted by the COT are simulation, ES(expert system) and numerical analysis. Although it is true that these methods can produce "sharp" results to a given problem, it is also well known that the results are sometimes too specific so that their general implication is severely limited. Obviously, although the OCT adopts the mathematical qualitative approach which can enjoy generality, sometimes it suffers from shortage of sharpness (specificity).

Ref. [1] quotes four research areas for the COT: Organization design, organizational learning, organization and information technology and organizational evolution and change.

According to this classification the OCT is mainly concerned with the organization design area. However, investigations of the design naturally induce investigation of organizational learning and organizational evolution and change.

Ref. [1] correctly points out that the COT has an advantage in dealing with a dynamic nature of an organization, change of organizational agent and change of an organizational environment. Furthermore, since it is strong in handling a complicated structure with many variables, the COT can yield a comprehensive view of a changing organization if the COT is used properly.

11.2 Garbage Can Model

The model of the OCT can be summarized in the following way:

The model of agents is a goal seeking system where an agent is assumed to behave in a rational way, i.e., their behaviors are assumed to be explained in terms of their goals. The organization is an open system which receives a (management information) input from its environment and while consuming management resources, yields a product to the environment as an output. An important assumption is that the management resources can be controlled although they may function as constraints. The organization itself has a goal—an organizational goal—which provides a measure to evaluate the outcome. In other words, an organization is to be managed to maintain an optimum (satisfactory) behavior adapting to the change of the environment. The optimality is measured by the organizational goal. The gap between the optimum state and the current state is a problem.

The required behavior is addressed by a hierarchically arranged internal structure. The hierarchy consists of three layers: Institutional layer, organizational layer and operational layer. The operational layer is responsible for the primary activity of an organization, transformation of management resources into products, and is assumed to be departmentalized to make the best use of professionalities of agents for the respective activities.

The change of the environment induces a problem to be solved and consequently a change of an internal goal, called global goal, which is an operationalization of the organizational goal. The modification of the global goal is done by the institutional level management. This is an adaptation activity of the OCT. The global goal is sent to the organizational level as a target for its tactical decision making activity.

The organizational level management is responsible for coordination of the departmentalized operational level activities. It is strongly desired that the organization structure is designed in a way that a successful coordination induces realization of the global goal.

The operational level goal seeking units are supposed to make decisions so as to optimize their respective goals. Their decisions determine directly the outcome of the organization.

This is a story of the OCT

The garbage can model, however, contradicts the above theory of the management behavior.

In the garbage can model there are two external inputs. One is called a choice opportunity. The opportunity can correspond to the concept of a problem situation [21], which demands a decision making activity. In the cybernetic model such an opportunity is supposed to induce an operational goal seeking problem but in the garbage can model, the opportunity cannot determine a problem directly.

The other input is a class of problems. The garbage can model assumes a pool of hidden problems which are fixed interpretations of alternative problem situations and the input is a flow of problems surfaced from the pool by chance. Some problems of the input are selected as immediate targets depending on the opportunity input while the others are saved as backlogs. The selection rule, which will be explained below, characterizes the property of an organization.

The garbage can model assumes an association relation between problems and choice opportunities. The relation is represented by a mapping from problems to opportunities.

In the garbage can model each problem is characterized by a difficulty level, which is measured by energy required to solve the problem. Then, a difficulty level is also induced for a choice opportunity, which is determined by the sum of the required energy to solve the problems attached to the opportunity. Decision making can be successful in the opportunity if available energy is sufficient to overcome the difficulty. This view that a problem and/or a decision making should be characterized by a difficulty measure does not exist in the OCT model formulation.

In the garbage can model not every agent is associated with a given choice opportunity. An association relation is also assumed as a mapping from agents to choices.

Capability of an agent is measured by the amount of his energy. If an agent is attached (or selected) to a choice opportunity, he is supposed to expend his entire energy to the opportunity in each period. If the sum of energies expended by the agents is larger than the required energy computed from its attached problems, it is said that a successful decision is made for the opportunity and its attached problems are assumed to be solved.

A choice opportunity input and problem inputs occur at a discretized interval. Problems are selected according to the occurrence.

It is important to notice that there can be backlogs of choices and problems. This situation is completely different from the arguments in the preceding chapters where a goal seeking problem is induced by a choice opportunity to be solved in a given interval T_{coord}. There is no concept of backlog in the previous chapters.

In the garbage can model an agent is an input/output system. Attachment of an agent to a choice opportunity is an input to him. He is supposed to willingly serve his task unconditionally when he is attached to it. The attachment is made by a rule which characterizes an organization.

The garbage can model does not have an explicitly stated organizational goal. The model says that the function of an organization is to make decisions, if possible, (using a standard procedure) for given opportunities and then problems attached to these opportunities are solved. It can be said then that an organization has an implicit goal which specifies how to handle the choice opportunities.

In summary decision making and problem solving are done when problems and agents are attached to choice opportunities or when they are thrown into garbage cans called choice opportunities. Problems and their solving activities are not related in a logical way.

11.3 Formulation of Garbage Can Model

The characteristics of the garbage can model are:

1. Problem situations and problems are assumed standardized.
2. Decision is assumed made by a standard procedure.
3. Professionality of an agent is explicitly appreciated.
4. Although Ref. [10] mentions hierarchy of an organization, no characterization of it is given .

Then, although the garbage can model is a comprehensive model of an organization, the most natural interpretation of the model is that it describes the operational level behaviors or represents behaviors of an organization as the operational level of the OCT. This section will formalize the garbage can model following this understanding.

In the cybernetic model management resources are assumed controllable and hence, are treated as parts of the decision variable or as constraints of a goal seeking problem. But when the cybernetic model is interpreted in terms of the garbage can model, the constraints by the management resources are considered to correspond to the problems of the garbage can model.

Obviously, constraints are problems for decision making and change dynamically following the environmental change.

Then, the following formulation assumes that the choice opportunity and the problems correspond to the global goal and the constraints input, respectively. The operational units correspond to agents whose model is an input output system. Fig. 11.1 shows the interpretation.

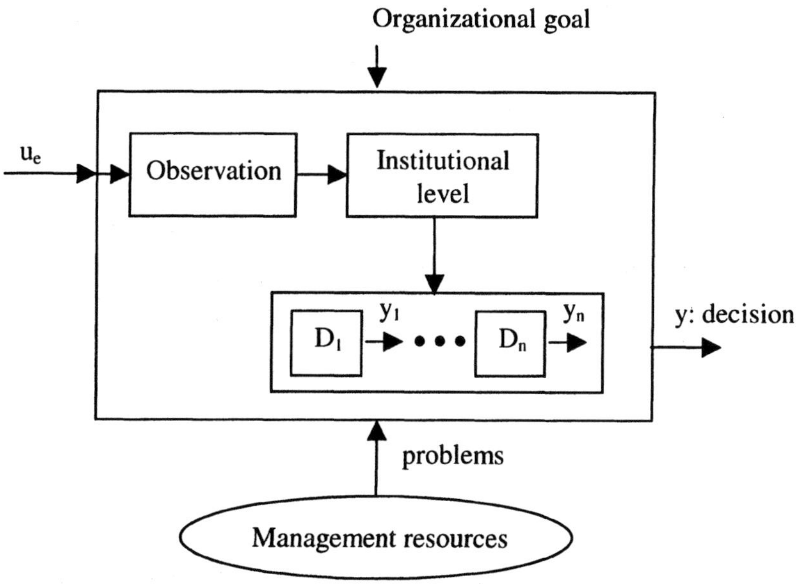

Figure 11.1 Garbage can model

u_e of Fig. 11.1 represents the environment. The institutional level produces a global goal or a choice opportunity G_{i_c} depending on u_e and the organizational goal. The world the garbage can address consists of D_i (agent), problems associated with management resources and decision y_i and y.

The garbage can model can be formalized ultimately as an automaton. That is,

$$\text{garbage can model} = \langle A, C, \delta \rangle \equiv M_{GC}$$

where

$$A = \text{choice opportunities} \times \{\text{problems}\},$$

C={(<backlog of choice opportunities>,<backlog of problems>, <expended energy>, <required energy>)}

and

$$\delta : C \times A \rightarrow C.$$

The precise definitions of A and C will be given below.

The following are basic assumptions to specify δ [10].

Assumption 1: Energy additivity assumption

Each choice requires as much of effective energy as the sum of all requirements of the several problems attached to it. The effective energy devoted to a choice is the sum of energies of the decision makers attached to that choice, deflated, in each time period, by the solution coefficient. As soon as the total effective energy that has been expended on a choice equals or exceed the requirements at a particular point in time, a decision is made.

Assumption 2: Energy allocation assumption

The energy of the each participant is allocated to no more than one choice during each time period. Each participant allocates his energy among the choices for which he is eligible to the one closest to decision, that is the one with the smallest energy deficit at the end of the previous time period in terms of the energies contributed by other participants.

Assumption 3: Problem allocation assumption

Each problem is attached to no more than one choice each time period, selected from those accessible by calculating the apparent energy deficit (in terms of the energy requirements of other problems) at the end of the previous time period and the selection of the choice closest to the decision.

Let us consider the normative model introduced in Chapter 9 as a prototype for the current formalization of the garbage can model.

Let the set of global goals or choice opportunities $G_{i_c} : Y \rightarrow Re$ be finite, i.e.,

$$i_c \in I_c = \{1,2,..,n_c\}.$$

Let V be a set of directives.

Let the set of constraints or problems $R^{i}_{p} : V \rightarrow \{0,1\}$ be finite, i.e.,

$$i_p \in I_p = \{1,..,n_p\}.$$

If $R^{i_p}(v)=1$, the directive v is said to satisfy the i_p-th problem (constraint) or the i-th constraint is not a problem for feasibility of v.

Let the index set of agents (operational units) be

$$I_d=\{1,..,n\}.$$

Then, two access relations, $RtoC:I_p \times I_c \rightarrow \{0,1\}$ and $DtoC:I_d \times I_c \rightarrow \{0,1\}$ can be defined.

Suppose a decision making opportunity is given as G_{i_c} where $i_c \in I_c$. Then,

$$RtoC(i_p,i_c)=1 \leftrightarrow (\exists v^*)(\max\{G_{i_c}(P^n(v))|v \in V\}=G_{i_c}(P^n(v^*)) \text{ and } R^{i_p}(v^*)=0).$$

In other words, the directive v^* is the most desirable for the goal G_{i_c} but its feasibility is hampered by the i-th constraint (v^* does not satisfy the constraint). Then, the constraint becomes a problem to be resolved for G_{i_c}.

Suppose a decision making opportunity is given as G_{i_c} where $i_c \in I_c$. Since $G_{i_c}:Y_1 \times .. \times Y_n \rightarrow Re$, a sensitivity measure can be defined for G_{i_c} with respect to y_j in the following way:

$$\partial G_{i_c}(y_1,..,y_n)/\partial y_j.$$

The above expression is a conceptual one because in general y_j is a time function. Let a threshold of the sensitivity be $a \in Re$. Then

$$DtoC(j,i_c)=1 \leftrightarrow |\partial G_{i_c}/\partial y_j| \geqq a.$$

In other words, if $DtoC(j,i_c)=1$, the activity of the j-th unit can affect the goal G_{i_c} in an effective way.

We will proceed with our arguments about the garbage can model assuming RtoC and DtoC are given a priori. That is, in general let an *access relation between problems and choice opportunities* be

$$RtoC:I_p \times I_c \rightarrow \{0,1\}.$$

If $RtoC(i_p,i_c)=1$, then the problem R^{i_p} can be attached to the choice opportunity G_{i_c}. R^{i_p} can be a problem of the decision making opportunity of G_{i_c} although R^{i_p} is not necessarily selected as a target problem of G_{i_c}.

Let an *access relation between operational units and choice opportunities* be

$DtoC:I_d \times I_c \rightarrow \{0,1\}$.

If $DtoC(i_d,i_c)=1$, then the operational unit D_{i_d} can be attached to the choice opportunity G_{i_c}. D_{i_d} can be selected to participate in solving problems of G_{i_c}.

Let an available energy for the operational units be

$eavd:I_d \rightarrow Re$.

That is, $eavd(i_d)$ is the amount of energy which the i_d -th operational unit can expend at each participation in problem solving.

Let the required energy for the problem R^{i_p} be

$ereqp:I_p \rightarrow Re$.

That is, $ereqp(i_p)$ is the amount of energy required to solve the problem R^{i_p}.

Then,

environmental condition of garbage can model= <RtoC, DtoC, eavd, ereqp, solcoeff>

where $solcoeff \in Re$ is a solution coefficient which specifies the efficiency factor of the energy expended.

Let

$\wp_p \subset \wp(I_p)$

be a partition of I_p. Then the input alphabet of $M_{GC} = \langle A,C,\delta \rangle$ is:

$A=I_c \times \wp_p$.

$(i_c,I_p') \in I_c \times \wp_p$ is an input of M_{GC} where $I_p' \subset I_p$. That is, at each period one choice opportunity i_c comes to M_{GC} to which more than one problems I_p' can be attached.

Let

$Eexpc= \{eexpc | eexpc:I_c \rightarrow Re\}$

where eexpc(i_c) is the *amount of energy which has been expended to the choice opportunity* G_{i_c}.

Let

$$Ereqc=\{ ereqc| ereqc:I_c \rightarrow Re\}$$

where ereqc(i_c) is the *amount of energy which is required to successfully make a decision for the choice opportunity* G_{i_c}.

Then, the state set C of M_{GC} is given by

$$C=Ereqc \times Eexpc \times \wp(I_c) \times \wp(I_p).$$

If (ereqc,eexpc,activechoice,activeprob)$\in$C,

$$activechoice \subset I_c$$

and

$$activeprob \subset I_p$$

represent backlogs of choices and problems to be processed.

Finally, let us construct δ of M_{GC}. Let

$$\delta((ereqc,eexpc,activechoice,activeprob),(i_c,i_{ps}))$$

$$= (ereqc',eexpc',activechoice',activeprob')$$

where $i_c \in I_c$ and $i_{ps} \subset I_p$. (ereqc',eexpc',activechoice',activeprob') will be specified below.

Let

$$tactivechoice=activechoice \cup \{i_c\}$$

and

$$tactiveprob=activeprob \cup i_{ps}.$$

tactivechoice and tactiveprob are choices and problems to be processed after the input (i,i_p) has occurred, respectively.

Let Raccess:$I_p \rightarrow \wp(I_c)$ be

$$\mathrm{Raccess}(i_p)=\begin{cases}\{i_c \mid \mathrm{RtoC}(i_p,i_c)=1\}\cap \mathrm{tactivechoice} & \text{if } i_p \in \mathrm{tactiveprob} \\ \phi & \text{otherwise}\end{cases}$$

$\mathrm{Raccess}(i_p)$ is the class of the opportunities to which i_p can be attached.
Let $\mathrm{Daccess}: I_d \rightarrow \wp(I_c)$ be

$$\mathrm{Daccess}(i_d)= \{i_c \mid \mathrm{DtoC}(i_d,i_c)=1\} \cap \mathrm{tactivechoice}\ .$$

$\mathrm{Daccess}(i_d)$ is the class of opportunities to which i_d can be attached.
Let us define $\mathrm{Rmap}: I_p \rightarrow I_c$ as follows:
If $\mathrm{Raccess}(i_p) \neq \phi$,

$$\mathrm{Rmap}(i_p)=i_c \leftrightarrow \min\{(\mathrm{ereqc}(i) - \mathrm{eexpc}(i)) \mid i \in \mathrm{Racces}(i_p)\}$$

$$=(\mathrm{ereqc}(i_c) - \mathrm{eexpc}(i_c)).$$

If $\mathrm{Raccess}(i_p)= \phi$,

$$\mathrm{Rmap}(i_p)=0.$$

Rmap is a rule which specifies how a problem is attached to a choice. The rule comes from Assumption 3.

Let us define $\mathrm{Dmap}: I_d \rightarrow I_c$ as follows:
If $\mathrm{Daccess}(i_d) \neq \phi$,

$$\mathrm{Dmap}(i_d)=i_c \leftrightarrow \min\{(\mathrm{ereqc}(i) - \mathrm{eexpc}(i)) \mid i \in \mathrm{Daccess}(i_d)\}$$

$$=(\mathrm{ereqc}(i_c) - \mathrm{eexpc}(i_c)).$$

If $\mathrm{Daccess}(i_d)= \phi$,

$$\mathrm{Dmap}(i_d)=0.$$

Dmap is also a rule which specifies how an operational unit is attached a choice. The rule comes from Assumption 2.

Then we have

$$ereqc'(i_c) = \begin{cases} sum(ereqc(i_p) \mid Rmap(i_p) = i_c\} & \text{if } i_c \in tactivechoice \\ 0 & \text{otherwise} \end{cases}$$

$$eexpc'(i_c)=eexpc(i_c)+sum\{eavd(i_d) * solcoeff|Dmap(i_d)=i_c\}.$$

Let

$$solC=\{i_c|ereqc'(i_c) \leqq eexpc'(i_c)\}$$

and

$$solP=\{i_p|Rmap(i_p)=i_c, i_c \in solC\}.$$

solC is the set of choices for which decisions are made and solP is the set of problems which are solved.

Finally, we have

$$activechoice'=tactivechoice - solC$$

and

$$activeprob'=tactiveprob - solP.$$

11.4 Implementation of Garbage Can Model

Sec. 11.3 formalized the garbage can model in the set theoretic way. Once a model is represented in the set theoretic terms, it is easy to represent the model in Prolog and the model can be analyzed using the computer simulation. This is another formal approach to the organization study. The semantic gap between a model in the set theory and its computer implementation is quite narrow when Prolog is used. Appendix of Chapter 11 shows a Prolog representation of the garbage can model. n_p=20, n_c=10 and n_d=10 are assumed following Ref. [10].

Although the choice opportunity input and the problem input are assumed to occur in a random way, they are expressed in two fixed sequences, choiceinput and probleminput following Ref. [10]. As the predicates choiceinput and probinput of the program show, G_{10} is the first opportunity input and $\wp_p$ is designed in the way that two problems occur simultaneously. At the first stage two problems R^8 and R^{20} occur. 0 in the input sequence indicates there is no input but since there are backlogs of choice opportunities and problems, model execution continues to process them.

Every function is expressed as a list in Prolog. For instance, eexpc([0,...,0]) specifies that the initial state of eexpc:$I_c \rightarrow Re$ takes the value 0 for every $i_c \in I_c$,that is, expended energy is equal to 0 at the initial stage.

Similarly, a set is also represented by a list. For instance, activechoice([]) implies that at the initial stage activechoice is an empty set, i.e., activechoice $= \phi$.

In order to examine behavior of the model it is necessary to execute the model changing the environmental condition. eavd(0,[0.1,..,1]) indicates that the eavd of Case 0 is [0.1,..,1].

RtoC and DtoC are also parameterized and the above convention is used for them. Ref. [10] claims that a hierarchical property can be studied by arranging 1 in DtoC matrix in a proper way. For instance, DtoC of Case 1 indicates that the higher the rank of an agent is, the more choice opportunities he can participate in.

Discussions about a real computer simulation and analysis of the results are beyond the scope of this book [10].

APPENDIX OF CHAPTER 11

```
/*garbage can model*/
/*******input sequence********/
/*choice opportunity input*/
choiceinput([10,7,9,5,2,3,4,1,6,8,0,0,0,0,0,
0,0,0,0,0]);
/*problem input*/
probinput([[8,20],[14,16],[6,7],[15,17],[2,
13],[11,19],[4,9],[3,12],[1,10],[5,18],0,0,0,0
,0,0,0,0,0,0]);

/****environment*****************/
/*energy available to operator*/
eavd([0.1,0.2,0.3,0.4,0.5,0.6,0.7,0.8,0.9,1]);
/*energy required for prob. to be solved*/
ereqp([1.1,1.1,1.1,1.1,1.1,1.1,1.1,1.1,1.1,1.
1,1.1,1.1,1.1,1.1,1.1,1.1,1.1,1.1,1.1,1.1]);
/*operator energy efficiency coeff*/
solcoeff([0.6,0.6,0.6,0.6,0.6,0.6,0.6,0.6,0.6,
0.6,0.6,0.6,0.6,0.6,0.6,0.6,0.6,0.6,0.6,0.6]);

/*prob. access structure:prob-choice*/
rtoc(
      [[1,1,1,1,1,1,1,1,1,1],
      [1,1,1,1,1,1,1,1,1,1],
      [0,1,1,1,1,1,1,1,1,1],
      [0,1,1,1,1,1,1,1,1,1],
      [0,0,1,1,1,1,1,1,1,1],
      [0,0,1,1,1,1,1,1,1,1],
      [0,0,0,1,1,1,1,1,1,1],
      [0,0,0,1,1,1,1,1,1,1],
      [0,0,0,0,1,1,1,1,1,1],
      [0,0,0,0,1,1,1,1,1,1],
      [0,0,0,0,0,1,1,1,1,1],
      [0,0,0,0,0,1,1,1,1,1],
      [0,0,0,0,0,0,1,1,1,1],
      [0,0,0,0,0,0,1,1,1,1],
      [0,0,0,0,0,0,0,1,1,1],
      [0,0,0,0,0,0,0,1,1,1],
      [0,0,0,0,0,0,0,0,1,1],
      [0,0,0,0,0,0,0,0,1,1],
      [0,0,0,0,0,0,0,0,0,1],
      [0,0,0,0,0,0,0,0,0,1]] );

/*operator access structure:operator-
choice*/
dtoc(
      [[1,1,1,1,1,1,1,1,1,1],
      [0,1,1,1,1,1,1,1,1,1],
      [0,0,1,1,1,1,1,1,1,1],
      [0,0,0,1,1,1,1,1,1,1],
      [0,0,0,0,1,1,1,1,1,1],
      [0,0,0,0,0,1,1,1,1,1],
      [0,0,0,0,0,0,1,1,1,1],
      [0,0,0,0,0,0,0,1,1,1],
      [0,0,0,0,0,0,0,0,1,1],
      [0,0,0,0,0,0,0,0,0,1]] );

/*main routine*/
garbage3():-
  function([eexpc,ereqc,activechoice,active
prob,solC,solP,rtoc,dtoc,eavd,ereqp,choice,
problem,createraccess,createdaccess,create
map,createereqc,createeexpc]),
/*simulation time span*/
   NTP:=20,
/*initialize state*/
   eexpc():=[0,0,0,0,0,0,0,0,0,0],
   ereqc():=[0,0,0,0,0,0,0,0,0,0],
   activechoice():=[],
   activeprob():=[],
   /*initialize auxiliary variable*/
   solC():=[],
   solP():=[],

/*delta(<state>,<input>)=(<next state>)*/
/*<state>=(eexpc,ereqc,activechoice,active
prob)*/
/*<input>=(choice,prob.)*/
/*Nc=dimension of choice=10*/
/*Np=dimension of prob=20*/
/*Nd=dimension of operator=10*/
/*process=implementation of delta*/
/*process(<Nc>,<Np>,<Nd>,<span>,<initia
l>)*/
   process(10,20,10,NTP,0),
   display_result();
display_result():-
   xwriteln(0,"END");

/*definition of Daccess*/
createdaccess(DtoC,TAC,Nd,Daccess):-
   Daccess:=constantlist(0,Nd),
   AC:=activechoice(),
   DC:=dtoc(),
   for(Id:=1;Id<=Nd;Id++)
   begin
```

```
      project(getvalue(DC,Id-1),Ics0,1,"1"),
      intersection(Ics0,AC,Ics),
       setvalue(Daccess,Id-1,Ics)
   end;

/*definition of Raccess*/
createraccess(RtoC,TAC,Np,Raccess):-
   Raccess:=constantlist(0,Np),
   AP:=activeprob(),
   AC:=activechoice(),
   RC:=rtoc(),
   for(Ip:=1;Ip<=Np;Ip++)
   begin
      if member(Ip,AP)   then

        project(getvalue(RC,Ip-1),Ics0,1,"1
        "),
         intersection(Ics0,AC,Ics)
      else
         Ics:=[]
      end,
         setvalue(Raccess,Ip-1,Ics)
      end;

mkmask([],M):-!;
mkmask([I|Is],M):-
   setvalue(M,I-1,0),
   mkmask(Is,M);

/*definition of Dmap and Rmap*/
createmap(Raccess,
   Estatusc,Np,Pmap):-
   Pmap:=constantlist(0,Np),
   assign(regI,0),
   assign(regP,Raccess),
   L:=strlen(Estatusc),
   repeat,
      regI(Ip),
       regP([X|Xs]),
      if X=[]   then
      else
         M:=constantlist(1000,L),
          mkmask(X,M),
          Estatusp:=Estatusc+M,
          min(Estatusp,Min,Pos),
          setvalue(Pmap,Ip,Pos+1)
      end,
      assign(regI,Ip+1),
      assign(regP,Xs),
         Xs=[],!,
   retract([regI,regP]);

/*definition of ereqc*/
createereqc([],Pmap,Nc,Ereqc):-!,
Ereqc:=constantlist(0,Nc);
createereqc(TAC,Pmap,Nc,Ereqc):-
   Ereqc:=constantlist(0,Nc),
   assign(regIc,TAC),
   repeat,
      regIc([Ic|Ics]),
      project(Pmap,Ips,Ic,"1"),
      project(ereqp(),Ips,Ep),
      setvalue(Ereqc,Ic-1,sum(Ep)),
      assign(regIc,Ics),
      Ics=[],!,
   retract([regIc]);

/*definition of eexoc*/
createeexpc([],Dmap,Coeff,Eavd,Nc,Eexpc
):-!,
   Eexpc:=constantlist(0,Nc);
createeexpc(TAC,Dmap,Coeff,Eavd,Nc,Ee
xpc):-
   Eexpc:=constantlist(0,Nc),
   assign(regIc,TAC),
   repeat,
      regIc([Ic|Ics]),
      project(Dmap,Ids,Ic,"1"),
      project(eavd(),Ids,Ed),

     setvalue(Eexpc,Ic-1,sum(Ed)*Coeff),
      assign(regIc,Ics),
      Ics=[],!,
   retract([regIc]);

solProb([],Pmap,[]):-!;
solProb([Ic|Ics],Pmap,Sol):-
   project(Pmap,Ips,Ic,"1"),
   solProb(Ics,Pmap,Sol1),
   append(Ips,Sol1,Sol);

choice(LT,C):-
   choiceinput(Cs),
   getvalue(Cs,LT,C);
problem(LT,P):-
   probinput(Ps),
   getvalue(Ps,LT,P);

/*implementation of delta*/
```

```
process(Nc,Np,Nd,NTP,NTP):-!;
process(Nc,Np,Nd,NTP,LT):-
/*get current state*/
    Seexpc:=eexpc(),
    Sereqc:=ereqc(),
    SAC:=activechoice(),
    SAP:=activeprob(),
    xwriteln(0,"eexpc=",Seexpc),
    xwriteln(0,"ereqc=",Sereqc),
    xwriteln(0,"activechoice=",SAC),
    xwriteln(0,"activeprob=",SAP),
/*get input*/
    C:=choice(LT),
    P:=problem(LT),
    xwriteln(0,"C=",C),
    xwriteln(0,"P=",P),
    /*set tactivechoice*/
    if C<>0   then
        inc(activechoice,[C])
    end,

    /*set tactiveprob*/
    if P<>0   then
        inc(activeprob,P)
    end,

/*get tactivechoice*/
    TAC:=activechoice(),
/*get Raccess:Ip -> power(Ic)*/
    Raccess:=createraccess(RtoC,TAC,Np),
/*get Daccess:Id -> power(Ic)*/
    Daccess:=createdaccess(DtoC,TAC,Nd),
/*Estatusc(ic)=ereqc(ic)-eexpc(ic)*/
    Estatusc:=Sereqc-Seexpc,
/*get Pmap*/
    Pmap:=createmap(Raccess,Estatusc,Np),
/*getDmap*/
    Dmap:=createmap(Daccess,Estatusc,Nd
),
/*get new ereqc*/
    Ereqc:=createereqc(TAC,Pmap,Nc),
    ereqc():=Ereqc,
    /*get new eexpc*/
    solcoeff(Cs),
    project(Cs,LT+1,Coeff),
    Eexpc1:=createeexpc(TAC,Dmap,Coeff,
Eavd,Nc),
    eexpc():=eexpc()+Eexpc1,
/*compute decided choice*/
    mkDec(Dec),
/*update solC*/
    inc(solC,Dec),
/*get new activechoice*/
    dec(activechoice,Dec),
/*compute solved prob*/
    solProb(Dec,Pmap,Sol),
/*update solP*/
    inc(solP,Sol),
/*get new avtiveprob*/
    dec(activeprob,Sol),
/*goto next stage*/
    process(Nc,Np,Nd,NTP,LT+1);

/*make decision*/
mkDec([]):-
    activechoice([]),!;
mkDec(Sol):-
/*get active choice*/
    L:=activechoice(),
    ERC:=ereqc(),
    EE:=eexpc(),
    assign(regSol,[]),
    NL:=strlen(L),
    for(IL:=0;IL<NL;IL++)
    begin
        C:=getvalue(L,IL),
        getvalue(ERC,C-1,ERCV),
        getvalue(EE,C-1,EEV),
        if EEV>=ERCV    then
            inc(regSol,[C]),
        else
        end
    end,
/*resolved choice*/
    regSol(Sol),
    retract([regSol]);

?-garbage3();
```

APPENDIX I GENERAL SYSTEMS APPROACH AND GENERAL SYSTEMS MODELS

AI.1 Introduction

There may be many opinions about the classical question what the general systems theory means. It is not easy to give a definite answer to the question. Fig. AI.1 shows a view of the general systems theory on which this book is based.

The general systems theory (GST) or in general any systems theory must start from the real world. Systems thinking, then, formalizes one aspect of the real world as an object which is called system. Let us call this formalization as the first system formalization and the object as a system object. Then theorization of the system object generates various individual systems theories. Typical examples are linear systems theory, automata theory, mathematical theory of economics and others.

The individual theories are further abstracted to find the essences of the system concept. The efforts yield systems philosophy and systems methodology. Bertalanffy's system philosophy is a typical example of them [19]. He got his idea of GST from his study on theoretical biology. Forrester's system dynamics [20] or Checkland's soft system methodology [21] may well be called examples of GST methodology generated from this kind of efforts.

There are other efforts made for the individual theories. Some people try to integrate the individual theories to yield grand theories which are usually called 2nd order systems theories. Boulding's hierarchy of the real world complexity can be considered 2nd order theory [22]. Laszlo's natural system concept [23] and Miller's living system concept [4] are typical grand theories.

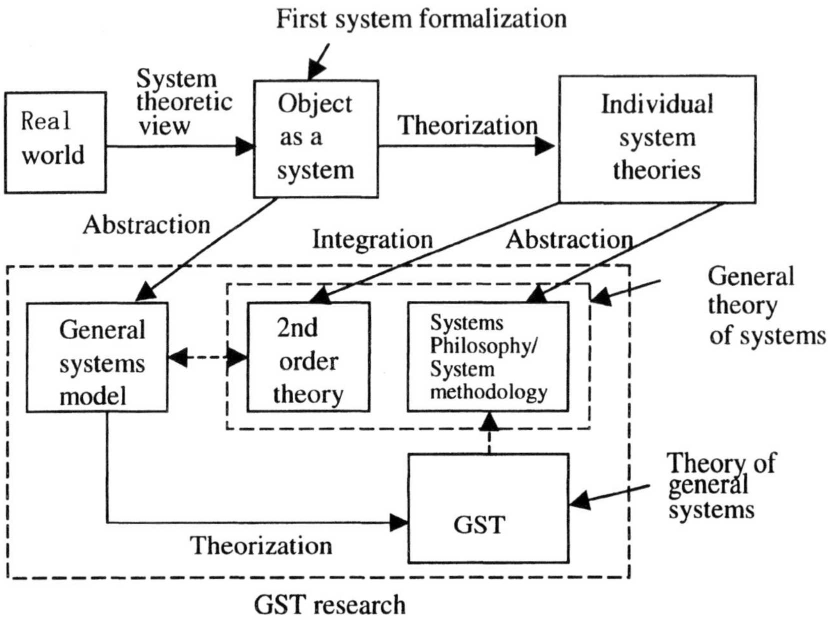

Figure AI.1 Overview of general systems theories

The natural system theory claims that its scope covers from an atomic system through society to the cosmic system while the living system concept does from a cell system through society to a supranational system.

Systems philosophy and systems methodology and 2nd order theories constitute a class of general system theories which may be called general theory of systems.

There is another kind of GST. By abstraction of the systems concept of the first formalization we can generate objects called general systems models. Then, we can build a general system theory by developing theories about the models. Usually the theories are formal ones and may well be called theories of general systems. This book is based on a theory of general systems.

There are not a few theories of general systems. Zadeh's fuzzy system theory may be one of the most influential ones of today [24]. Mesarovic developed a formal general systems theory[2]. Kir has a formal theory [25]. Wymore developed a formal theory for systems engineering [26]. Pichler's CAST (computer aided system theory) may be called a theory of a general system if we understand that a computer implementation is a type of general system representation [27]. Wu's pansystem is a formal GST [28].

AI.2 Development Methodology and General Systems Model

GST development of this book is based on the famous thesis of Boulding, that is, there is an ideal level of abstraction between a generalized nonsense and a meaningless specialization [22]. There are two ways to reach the level. One way is by generalization of the meaningless specialization. The other is by specialization of the generalized nonsense. This book followed the latter approach, that is, we start from a highly abstract general systems models and search for a convenient level to generate meaningful results by specialization of the models.

There are three typical general systems models for our approach, general input output model, goal seeking model and hierarchical model.

Our specialization process or theorizations process is based on the following principles:

(i) Characterization
(ii) Categorization
(iii) Model theoretic representation
(iv) Universal representation

Our approach is certainly a top down approach. In our top down approach an abstract object is specialized step by step introducing newly created concepts. The concepts must be systems theory oriented. The system of the new concepts is called GST characterization of the object [2].

It is our standpoint that GST is a meta theory which is complementary to individual theories. Then, GST must present a map of the individual theories showing relationships among them. The map is to clarify what kind of systems objects belong to a theory. Relationship representation and object definition can be conveniently treated by the category theory of mathematics. The categorical approach is then indispensable to our GST development [2].

We want GST to be a language for systems engineering. This has been one of the goals of GST study since its birth. In order to realize the goal we not only adopt a formal approach but also try to present concepts and results of GST in the model theoretic form of mathematics [29].

Universality of GST has two meanings. First, we want GST to provide a universal platform for individual systems theories [2]. GST must be a general theory.

The second universality means that GST must reveal essences of system models or concepts and hence constructive proofs are desired for proving theoretical facts of GST. This idea of universality comes from the concept of free algebra.

The subsequent sections will demonstrate how a mathematical GST can be developed by applying the above principles to the general systems models.

AI.3 Characterization

The characterization approach formalizes fundamental properties of a target object as general systems theoretic concepts. Specialization of an abstract model of GST is done using the concepts so the behavior or the nature of the target object can be satisfactorily explained in the framework of the specialized model. In this case there are three problems to be solved.

(i) Construction of a starting general system model.
(ii) Creation of general system theoretic concepts.
(iii) Explanation of the target system behavior applying the created concepts to the general system model.

Let us demonstrate the characterization approach using the linear systems theory as an example. The linear systems theory is a most deeply developed system theory so that it can be a good target for the characterization approach.

The target object of the linear systems theory is described by the following differential equation:

$$dz/dt = Fz+Gx$$

$$y=Hz$$

where x, y and z are input vector, output vector and state vector, respectively. F, G and H are constant matrices of appropriate dimensions.

The starting general model of GST is a general input output model which is given as follows: Let X and Y be arbitrary sets called input set and output set, respectively. Then a relation S

$$S \subset X \times Y$$

is called general input output system.

Let us consider a specialization process of the general input output system.

The first step is to specialize the general model as a dynamic model by introducing the time concept.

Let

A= set of input alphabet

B = set of output alphabet

$$T=[0, \infty)$$

T can be, in general, any set with a linear order. For the sake of simplicity, however, we assume that T is the right hand side of the real line as defined above. T is called time set.

Then, X and Y are structured as:

$$X \subset A^T$$

$$Y \subset B^T.$$

That is, X and Y are classes of time functions on A and B, respectively.

A time system which is an abstract model of a dynamic system is given by:

$$S \subset A^T \times B^T.$$

In order to give more descriptive power to the time system model general operations on time functions are introduced. These operations are described in Chap. 5 for the case of discrete system. Let

$$T^t = [0, t),$$

$$T_{ts} = [t, s),$$

$$T_t = [t, \infty),$$

$$\underline{T}^t = [0, t],$$

$$\underline{T}_{ts} = [t, s],$$

$$x^t = x|T^t,$$

$$x_t = x|T_t,$$

$$x_{ts} = x|T_{ts},$$

$$\underline{x}^t = x|\underline{T}^t,$$

$$\underline{x}_{ts} = x|\underline{T}_{ts}.$$

Then, the concatenation operation is introduced as follows:

$$x'' = x^t . x'_t$$

where

$$x''(s) = \begin{cases} x^t(s) \text{ if } s<t \\ x'_t(s) \text{ if } s \geq t \end{cases}$$

The concatenation operation is the most basic but powerful one in GST.

An initial response function of a time system S is introduced as the first created concept. Let C_0 be a set called initial state. Then, an initial response function

$$P_0 : C_0 \times X \rightarrow Y$$

of S is given by

$$(x, y) \in S \leftrightarrow (\exists\ c)(y = P_0(c, x)).$$

As for the initial response function concept we have the following fact:

Theorem AI.1 [2]
Every time system has an initial response function.

Let us introduce the causality concept in the domain of GST [2]. An initial response function $P_0 : C_0 \times X \rightarrow Y$ is called causal if and only if for any $c \in C_0$ the following holds: For any x and x' and t

$$x^t = x'^t \rightarrow P_0(c,x)|\underline{T}^t = P_0(c,x')|\underline{T}^t .$$

Then, a time system S is called *causal* if and only if S has a causal initial response function.

The causality can be characterized in two ways in GST. Let us adopt the simpler one where the past determinacy concept is used.

Let us say that a time systems S is past determined from s if and only if the following two conditions are satisfied:

(i) $(x,y)|T^s=(x',y')|T^s$ and $x^t=x'^t \rightarrow y^t=y'^t$ for $s \leqq t$.

(ii) $(x^s.x'_s, y^s.y'_s) \in S$ for any $(x,y),(x',y') \in S$.

It can be shown that the target object of the linear system theory possesses the past determinacy property.

Then, the causality can be characterized by the following result:

Theorem AI.2 [2]

S_s is causal if S is past-determined from s where

$$S_s = S|T_s = \{(x,y)|T_s \mid (x,y) \in S\}.$$

Next, let us introduce the concept of state space representation for GST on the causality concept [2]. For a time system $S \subset X \times Y$ and for a set C let

$$\underline{\phi} = \{\phi_{st} \mid \phi_{st}: C \times X_{st} \rightarrow C \text{ and } s,t \in T \text{ and } s \leqq t\}$$

and

$$\underline{\mu} = \{\mu_t \mid \mu_t: C \times A \rightarrow B \text{ and } t \in T\}.$$

Then $<\underline{\phi},\underline{\mu}>$ is called state space representation of S if and only if the following conditions are satisfied:

(i) $\underline{\phi}$ satisfies

(α) $\phi_{su}(c,x_{su}) = \phi_{tu}(\phi_{st}(c,x_{st}),x_{tu})$ (composition property)

where $s \leqq t \leqq u$

(β) $\phi_{tt}(c,x_{tt}) = c$ (consistency property).

(ii) $(x,y) \in S \leftrightarrow (\exists\, c \in C)(\forall\, t)(y(t)=\mu_t(\phi_{0t}(c,x^t),x(t)))$.

ϕ_{st} and μ_t are called *state transition function* and *output function*, respectively, where C is a state set.

Then, we have a characterization theorem of the state space representation concept.

Theorem AI.3 [2]

A time system S is causal if and only if it has a state space representation.

The proof of the only if part is difficult.

Let a response function at t be defined as:

$$P_t(c,x_t)(s)=\mu_s(\phi_{ts}(c,x_{ts}),x_t(s)) \quad \text{where } t \leqq s.$$

Let us introduce the concept of stationarity for GST [2].

Let σ^t be a shift operator for a time function which is specified as:

$$\sigma^t(x_{uv})(s)=x_{uv}(s-t) \quad \text{where } u+t \leqq s \leqq v+t.$$

That is, σ^t shifts x_{uv} by t parallel to the time axis. The shift operation is another basic powerful operation of GST.

Let

$$\lambda^t(x)=\sigma^{-t}(x|T_t).$$

Then, a state space representation $<\underline{\phi},\underline{\mu}>$ is called *time invariant* if and only if the following conditions are satisfied:

(i) $P_t(c,x_t)=\sigma^t[P_0(c,\sigma^{-t}(x_t))]$.

(ii) $\phi_{ts}(c,x_{ts})=\phi_{0u}(c,\sigma^{-t}(x_{ts}))$ where $u=s-t$.

(iii) $\lambda^t(X)=X$.

The stationarity concept of GST can be formalized in the following way:

A time system S is called *stationary* if and only if it has a time invariant state space representation.

A characterization theorem of the stationarity concept is given by:

Theorem AI.4 [2]
A time system S is stationary if and only if for any t

$$\lambda^t(S) \subset S.$$

As the final building block of concepts let us introduce the linearity for GST. Suppose the input and output alphabets A and B are linear spaces over the same field F. Then, A^T, B^T and $A^T \times B^T$ are linear spaces over F by the conventional definitions.

The linearity concept for GST is given by: A time system S is *linear* if and only if S is a linear subspace of $A^T \times B^T$, i.e.,

$$(\forall s, s' \in S)(\forall \alpha, \beta \in F)((\alpha s + \beta s') \in S).$$

Finally, let us characterize the target object of the linear system theory introducing the concept of basic linear system of GST. A time system S is called *basic linear system* if and only if it satisfies the following conditions:

(i) S is linear.
(ii) S is causal.
(iii) S is stationary.
(iv) S has a finite dimensional system core, i.e., dimS(0)} <∞
where $S(0) = \{y|(0,y) \in S\}$.

Using the basic linear system concept we can explain or characterize properties of the target object. The following is an example.

Theorem AI.5 (representation theorem) [2]
Suppose S is a basic linear system. Then,

(i) S has a canonical state space representation $(\underline{\phi}^*, \underline{\mu}^*)$ *which is time invariant.*

(ii) if $(\underline{\phi}, \underline{\mu})$ *is a reduced time invariant state space representation of S, then* $(\underline{\phi}, \underline{\mu})$ *is isomorphic to* $(\underline{\phi}^*, \underline{\mu}^*)$*, that is, we have commutative diagrams of Fig. AI.2 where h is a linear one-to-one correspondence.*

The above theorem says that a basic linear system has essentially a unique time invariant state space representation. This is not a trivial fact.

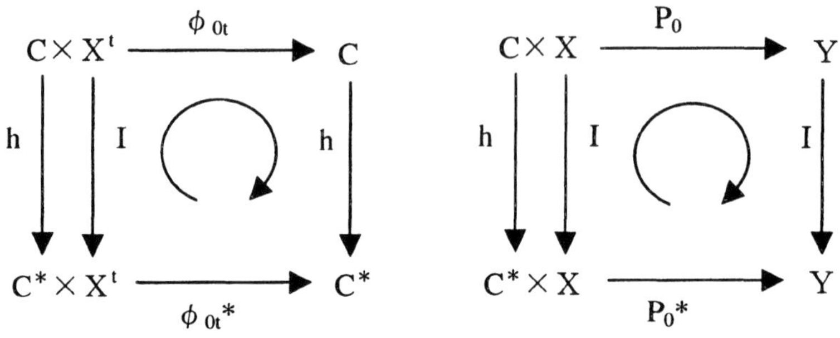

Figure AI.2 Uniqueness of canonical representation

AI.4 Categorization

We mentioned that the categorical approach is indispensable to GST as a meta theory. In this approach a system theoretic object is represented as an object of a category. In general a system theoretic object has a complicated structure like a state space representation. It is necessary but not easy to define a morphism meaningfully reflecting the structure of objects for the definition of a category. In our case a structural similarity is used for the morphism which will be discussed later.

In systems theory there are transformations from a class of system theoretic objects to another class which play important roles. An example is a transformation of a time system to its state space representation. It is natural, then, that in the categorical approach a transformation is formalized as a functor. Furthermore in some occasions more than one transformations are defined from the same domain to the same co-domain. In this case the relationships between the transformations can be formalized as natural transformations. In this way we can construct a meta theory of GST in the categorical approach.

Let us demonstrate the categorical approach for the goal seeking model which is one of the fundamental models of GST [2].

Fig. AI.3 shows a simplified structure of the goal seeking model. (Refer to Fig. 2.2.)

The model consists of two components, decision maker D and process P: The process P is an input and output model which has two inputs, decision variable $m \in M$ and external input $u \in U$. The output is $y \in Y$. The decision maker, using a feedback information about y and a feed forward information about u, selects an appropriate value for m. The selection is guided by the goal G and the decision principle ϕ.

Let us formulate the goal seeking model.

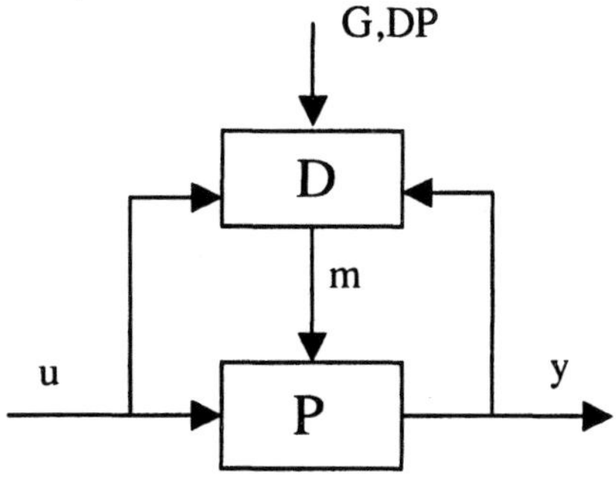

Figure AI.3 Goal seeking system

Let

$P:M \times U \to Y$; process.

$G:M \times U \times Y \to Re$; goal.

The goal G is an evaluation of the behavior of P. Then a goal seeking problem of GSP is given by

GSP=<M, U, Y, P, G>.

Let us introduce a reduced GSP for the sake of notational convenience. Let $g : M \times U \to Re$

g(m, u)=G(m, u, P(m, u)).

Then, a reduced GSP, RGSP is

RGSP=<M, U, g>.

Let us define a decision principle which characterizes the basic nature of a goal seeking system.

Let

<u>RGSP</u>= class of RGSP.

Let

$OALT=<M, \leqq> =$ ordered alternative structure of M

where $\leqq \subset M \times M$ is an order relation.

Let

<u>OALT</u>= class of ordered alternative structure.

Then, a decision principle DP is given as:

DP:**<u>RGSP</u>** →**<u>OALT</u>**

such that

$DP(<M, U, g>)=<M, \leqq>$.

A typical example of a decision principle is the max-min principle. That is,

$\text{max-min}(<M, U, g>)=<M, \leqq_{mm}>$

where

$m \leqq_{mm} m' \leftrightarrow \min_u g(m, u) \leqq \min_u g(m', u)$.

A goal seeking system is then given by

goal seeking system=<RGSP,DP >.

In order to define a category of goal seeking systems, let us introduce the concept of similarity morphism. Let <M, U, g> and <M', U', g'> be elements of **<u>RGSP</u>**. Let $h= <h_1, h_2, h_3>$ be $h_1 : M \to M'$, $h_2 : U \to U'$, and $h_3 : Re \to Re$. Then, h is called a *similarity morphism*,

$h: <M, U, g> \to <M', U', g'>$

if the fol lowing conditions are satisfied:

(i) The commutative diagram of Fig. AI.4 holds:

(ii) h_3 is a strict order homomorphism, i.e. ,

$r \leqq r' \to h_3(r) \leqq h_3(r')$ and $r< r' \to h_3(r)< h_3(r')$.

Then, we can define a category of RGSP, **<u>RGSP</u>** as follows:

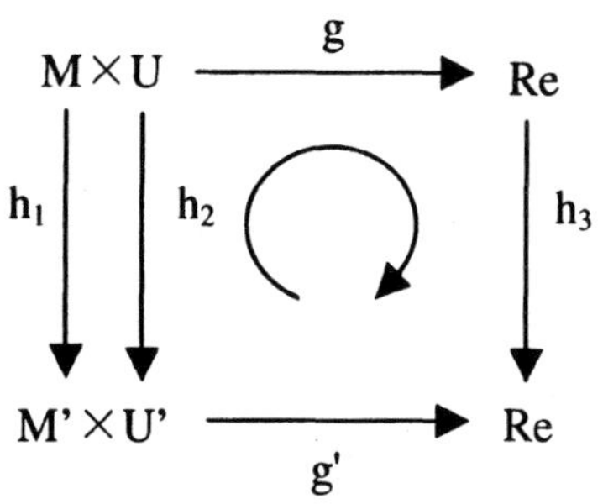

Figure AI.4 Morphism of goal seeking systems

Ob(**RGSP**) = class of RGSPs,

$\mathrm{Mor}(<M, U, g>, <M', U', g'>) = \{h \mid h = \{\text{similarity morphism from } <M, U, g> \text{ to } <M', U', g'>\}$.

It is easy to prove that **RGSP** satisfies the axioms of a category.

In the same way, we can define a category of OALT, **OALT** as follows:

Ob(**OALT**) = class of OALTs.

$\mathrm{Mor}(<M, \leqq>, <M', \leqq'>) = \{h \mid h: M \rightarrow M' \text{ is a strict order homomorphism}\}$

where

$m \leqq m' \rightarrow h(m) \leqq h(m')$ and $m < m' \rightarrow h(m) < h(m')$.

It is also easy to prove that **OALT** is a category.

Let us introduce a functor F_p, Pareto functor as follows:

$F_p(<M, U, g>) = <M, \leqq_p>$

where

$m \leqq_p m' \leftrightarrow (\forall u)(g(m, u) \leqq g(m', u))$

and

$F_p(\langle h_1, h_2, h_3\rangle)=h_1$ for $\langle h_1, h_2, h_3\rangle \in$ Mor(**RGSP**).

Then, it can be proved that F_p is a functor from **RGSP** to **OALT**.

Finally, let us define the concept of rational decision principle functor. Let F:**RGSP** →**OALT** be a functor. Then F is called a rational decision principle functor if and only if it satisfies the following conditions:

(i) $F(\langle M, U, g\rangle)=\langle M, \leqq\rangle$.

(ii) $F(\langle h_1, h_2. h_3\rangle)=h_1$.

(iii) There exists a natural transformation $\eta : F_p \rightarrow F$ such that

$\eta_\Delta(m)=m$ for any $\Delta=\langle M, U, g\rangle$ and $m \in M$, and the commutative diagram of Fig. AI.5 holds for $h:\Delta \rightarrow \Delta'$.

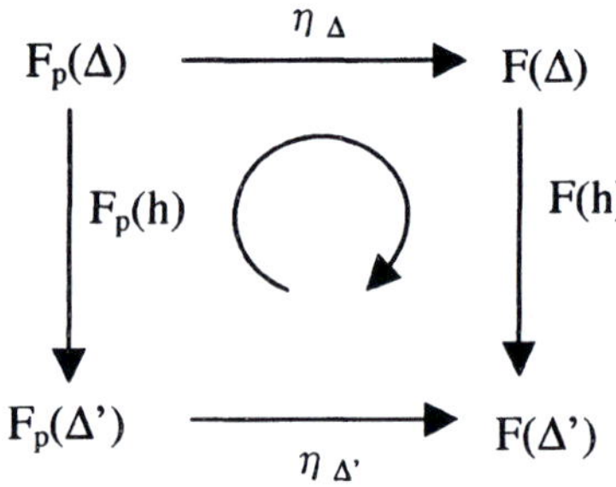

Figure AI.5 Natural transformation

A rational decision principle functor is a GST characterization of a rational decision. The following is a typical meta theoretic result of GST.

Theorem AI.6 [2]

*A decision principle DP: Ob(**RGSP**) → Ob(**OALT**) is the max-min decision principle if and only if it has an associate rational decision principle functor F : **RGSP**→**OALT** such that F(Δ) is continuous with respect to Δ.*

AI.5 Universality

As mentioned in AI.2, there are two types of universality. It is clear that GST arguments satisfy the first type of universality.

Let us consider the second type, the free construction type universality. We have demonstrated that many GST concepts are introduced on the specialization process. If a concept is created constructively, we say it is a free construction concept.

The free construction is well dealt with by the categorical theory. Let us recognize the specialization operation as a transformation from a less structured class to a more structured class. Typically, an element of the less structured class is a functional specification (input output specification) of a system while its corresponding element of the more structured one is its state space representation.

Let us consider Fig. AI.6 where the specialization is formulated by a realization functor while the abstraction from the more structured class to the less structured class is by a forgetful functor.

Let X be, for instance, a functional specification of a system object. Let U_x be a state space representation of X. Then, the freeness of U_x is characterized by a universal solution: Let G be a forgetful functor or $G(U_x)$ is a functional specification corresponding to U_x which may or may not be equal to X. Let $\rho_x \in Mor(G(U_x), X)$. Let S be an arbitrary state space representation. Let G(S) be its functional specification. Suppose $h \in Mor(G(S), X)$ is any morphism. Then, if there exists a unique morphism $h^* : S \rightarrow U_x$ such that $G(h^*) \rho_x = h$ holds, (U_x, ρ_x) is called co-universal solution with respect to G.

There are several famous free constructions. The following is a list of them.

1. sequential function $\rightarrow$
   state space representation by Nerode realization
2. input output model $\rightarrow$
   state space representation by Mesarovic realization
3. axiomatic system of basic linear system $\rightarrow$
   canonical state space realization
4. consistent set Σ of 1st order language $\rightarrow$
   canonical structure of model of Σ
5. game situation $\rightarrow$
   implementation theory of economics
6. algebraic structure $\rightarrow$
   free algebra

A free algebra is a classical example of the free construction.

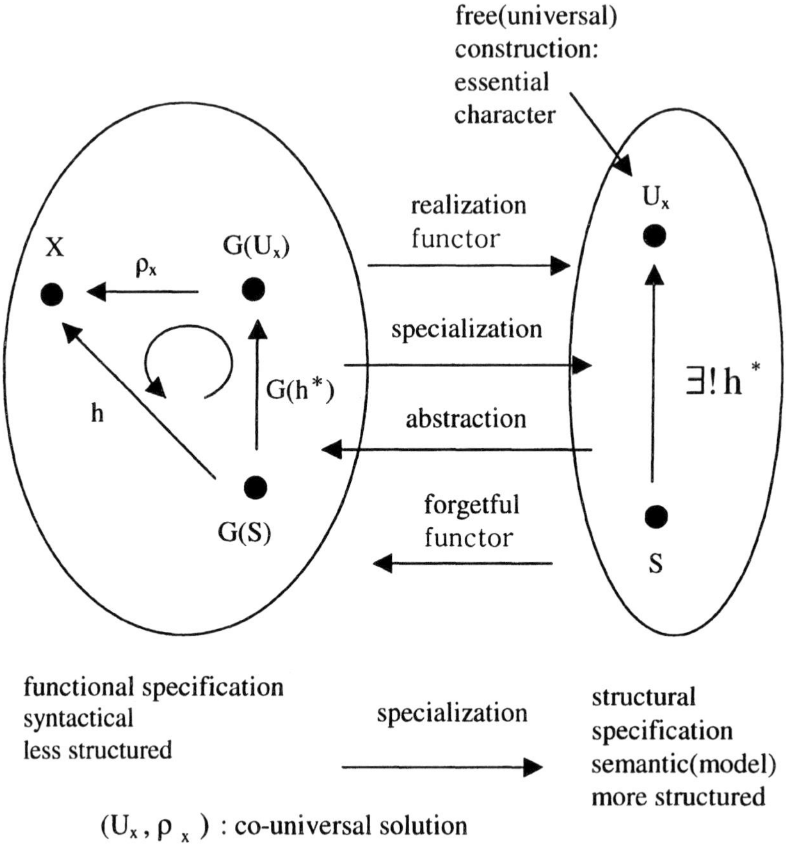

Figure AI.6 Universality of GST

Let us demonstrate the free construction by using Mesarovic realization as an example.

Let

$$S \subset X \times Y$$

be a time system. In order to get a state space representation of S, we have to create a state space. The free construction requires that the state space should be constructed by a direct use of S.

Let

$$C_t^* = \{c_t \mid c_t \subset S_t \text{ and } c_t : X_t \rightarrow Y_t\}.$$

C_t^* will be called pre-state space object at t. Notice that C_t^* is a constructive definition in terms of S. Then, based on C_t^* we can construct a pre-dynamical system representation of S which can be transformed into a state space representation.

Let $P_t : C_t^* \times X_t \rightarrow Y_t$ such that

$$P_t(c_t, x_t)=c_t(x_t).$$

Let $\phi_{ts}^* : C_t^* \times X_{ts} \rightarrow C_s^*$ such that

$$\phi_{ts}^* (c_t, x_{ts})=c'_s \leftrightarrow c'_s (x'_s)=c_t(x_{ts} \cdot x'_s)|T_s.$$

Then, <{ P_t^* },{ϕ_{ts}}> is called free construction of a pre-dynamical system representation of S.

Actually its freeness can be expressed as a universal solution. Let **disc-S** be the discrete category of time systems, that is

Ob(**disc-S**) = class of time systems

$$\mathrm{Mor}(S,S')=\begin{cases}(id,\ id) \text{ if } S=S' \\ \phi \text{ if } S \neq S'\end{cases}$$

The discreteness is essential because we want to demonstrate the special role of the free construction among state space representations of the same functional specification.

Let **PDSR** be a category of PDSRs where

Ob(**PDSR**) = class of PDSRs,

$$\mathrm{Mor}(<\underline{P},\underline{\phi}>, <\underline{P}',\underline{\phi}'>) \ni \underline{h}$$

where $\underline{P}$={Pt},$\underline{\phi}$={ϕ_{ts}} and $\underline{h}=\{h_t | h_t : C_t \rightarrow C'_t \text{ and } t \in T\}$ which satisfies the commutative diagrams of Fig. AI.7.

A morphism of **PDSR** is a similarity morphism.

Let G : **PDSR**→**disc-S** be a forgetful functor where

$$G(<\underline{P},\underline{\phi}>)=S$$

where <$\underline{P}$,$\underline{\phi}$> is a pre-dynamical system representation of S and

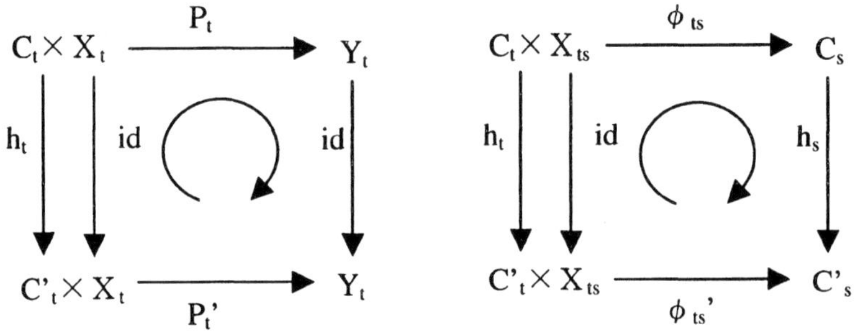

Figure AI.7 Mesarovic realization

$G(\underline{h}) = (id, id)$.

Then, we have,

Theorem AI.7 [2]

Let $\langle \underline{P}^*, \underline{\phi}^* \rangle$ be the free PDSR of S. Then, $(\langle \underline{P}^*, \underline{\phi}^* \rangle, \langle id, id \rangle)$ is a co-universal solution with respect to G.

APPENDIX II MIN-MAX THEOREM

AII.1 Introduction

Maximization or minimization is the most important operation for management since it is a fundamental of a rational activity. These operations are treated as an optimization in this book. This appendix will discuss the min-max theorem as a theoretical base of an optimization operation.

Since an organization is a dynamical system, a variable of its model is a time function or a real function on a time set (Refer to Appendix I). A space of time functions is in general an infinite dimensional linear space. In order to avoid mathematical difficulties due to the infinite dimensionality this appendix assumes finite dimensional linear spaces or Euclidean spaces for variables of a model. This assumption is acceptable because when a dynamical system is modeled and processed on a computer a time function is necessarily represented as a discrete function over a finite interval or as a finite dimensional vector. Nowadays, computation of optimization is almost always carried out on a computer in practice.

AII.2 Preliminary Concepts

This section will introduce several basic concepts necessary for this appendix.

Definition AII.1 Euclidean space

An n-dimensional real Euclidean space E_n is:

$$E_n = Re^n$$

which is a linear vector space with respect to the term wise operations of addition and scalar multiplication, i.e.,

$$a(r_1,..,r_n)+b(r_1',..,r_n')=(ar_1+br_1',..,ar_n+br_n')$$

where $r_i,r_i',a,b \in Re$.

Definition AII.2 Norm of E_n

A norm

$$\| \ \| : E_n \rightarrow Re$$

is an abstraction of the absolute value, which satisfies the following axioms:

(i) $\| x \| \geqq 0$ and $\| x \| = 0 \leftrightarrow x=0$,

(ii) $\| rx \| = |r| \, \| x \|$ where $r \in Re$,

(iii) $\| x+y \| \leqq \| x \| + \| y \|$.

A typical example of norm $\| x \|$ of $x=(r_1,..,r_n)$ is given by:

$$\| x \|^2 = r_1^2 + \cdots + r_n^2.$$

This appendix assumes the above form of norm unless explicitly mentioned otherwise.

Definition AII.3 Convex set

A subset X of a Euclidean space is called convex if and only if the following holds:

$$(\forall x_1,x_2 \in X)(\forall r \in Re)(0 \leqq r \leqq 1 \rightarrow rx_1+(1-r)x_2 \in X).$$

Definition AII.4 Convex function

Let X be a convex subset of a Euclidean space. A function $f:X \rightarrow Re$ is called a convex function if and only if the following holds:

$$(\forall x_1,x_2 \in X)(\forall r)(0 \leqq r \leqq 1 \rightarrow f(rx_1+(1-r)x_2) \leqq rf(x_1)+(1-r)f(x_2)).$$

Definition AII.5 Concave function

Let X be a convex subset of a Euclidean space. A function $f:X \to Re$ is called a concave function if and only if the following holds:

$$(\forall x_1,x_2 \in X)(\forall r)(0 \leqq r \leqq 1 \to f(rx_1+(1-r)x_2) \geqq rf(x_1)+(1-r)f(x_2)).$$

It should be noticed that a linear function $f:X \to Re$, which satisfies the following relation

$$f(ax+by)=af(x)+bf(y),$$

is a convex as well as a concave.

Definition AII.6 Distance between subsets

Let S_1 and S_2 be subsets of the n-dimensional Euclidean space E_n. Then, let the distance between them be:

$$d(S_1,S_2)=\inf_{(x,y)} \| y-x \|$$

where $x \in S_1$,$y \in S_2$.

Definition AII.7 Hyper plane

Let a subset HP of E_n be: For some $p=(p_1,..,p_n) \in E_n$ and $r \in Re$

$$HP=\{(x_1,..,x_n)| \Sigma_i p_i x_i=r \}.$$

Then HP is called hyper plane of E_n.

In the subsequent discussions the inner product of points x and x' of a Euclidean space will be denoted by xx'.

Definition AII.8 Convex hull

Let $X=\{x_1,..,x_n\}$ be a finite subset of a linear space. Then

$$X^h=\{x|x=a_1x_1+\cdots+a_nx_n, 0 \leqq a_i \leqq 1, a_1+\cdots+a_n=1\}$$

is called convex hull of X.

AII.3 Optimization and Min-max Theorem

The final target of this section is to prove the following theorem.

Theorem (min-max theorem)

Let X and Y be convex closed bounded subsets of Euclidean spaces. Let $f: X \times Y \rightarrow Re$ be a continuous function on $X \times Y$. Suppose f is a concave with respect to x for each y and a convex with respect to y for each x. Then, the following holds:

$$min_y \, max_x f(x,y) = max_x \, min_y f(x,y).$$

In order to show a relationship between the min-max theorem and an optimization problem let us consider the following simple optimization problem.

Let X be a subset of a Euclidean space.

Let $h: X \rightarrow Re$ and $g: X \rightarrow E_n$.

Let an optimization problem in consideration be:

$$h(x) \rightarrow \max$$

subject to

$$g(x)=0.$$

Let the Lagrangean function f(x,y) of the above problem be:

$$f(x,y)=h(x)+yg(x)$$

where $y \in E_n \equiv Y$.

Let

$$X_1=\{x | g(x)=0, \, x \in X\} \neq \phi.$$

Then, let us observe that

$$\sup_x \inf_y f(x,y)=\sup_{X_1} h(x)$$

because for any $x \notin X_1$ we have $\inf_y f(x,y)= -\infty$.

We will show that

$$\min_y \max_x f(x,y)=\max_x \min_y f(x,y)$$

is equivalent to the following relation:

There exists $(x_0, y_0) \in X \times Y$ such that

$$f(x,y_0) \leqq f(x_0,y_0) \leqq f(x_0,y). \qquad \text{(A)}$$

Suppose

$$\min_y \max_x f(x,y) = \max_x \min_y f(x,y)$$

is true. Let

$$\min_y \max_x f(x,y) = \max_x f(x,y_0) \geqq f(x,y_0)$$

and

$$\max_x \min_y f(x,y) = \min_y f(x_0,y) \leqq f(x_0,y).$$

Since

$$\min_y \max_x f(x,y) = \max_x \min_y f(x,y)$$

is true, we have

$$f(x,y_0) \leqq f(x_0,y)$$

for any $x \in X$ and $y \in Y$. Then if $x = x_0$,

$$f(x_0,y_0) \leqq f(x_0,y)$$

holds. If $y = y_0$,

$$f(x,y_0) \leqq f(x_0,y_0)$$

holds. Consequently, we have

$$f(x,y_0) \leqq f(x_0,y_0) \leqq f(x_0,y).$$

Conversely, suppose (A) holds. Then

$$\min_y \max_x f(x,y) \leqq \max_x f(x,y_0)$$

$$=f(x_0,y_0)=\min_y f(x_0,y)$$

$$\leqq \max_x \min_y f(x,y).$$

In general,

$$\min_y \max_x f(x,y) \geqq \max_x \min_y f(x,y). \text{ (Refer to Lem. 7.2)}$$

As a matter fact,

$$\min_y f(x,y) \leqq f(x,y) \leqq \max_x f(x,y)$$

is true for any x and y. Or

$$\min_y f(x,y) \leqq \max_x f(x,y)$$

is true for any x and y. Then

$$\max_x \min_y f(x,y) \leqq \max_x f(x,y)$$

is true for any y or

$$\max_x \min_y f(x,y) \leqq \min_y \max_x f(x,y).$$

Then,we have

$$\min_y \max_x f(x,y)=\min_y f(x_0,y)=\max_x \min_y f(x,y).$$

Then, x_0 is an optimum solution because

$$h(x_0) \geqq \min_y f(x_0,y) \text{ (because } h(x_0)=f(x_0,0))$$

$$=\max_x \min_y f(x,y)= \max_{x_1} h(x).$$

Since $x_0 \in X_1$, x_0 is an optimum solution. That is, x_0 is given by maximization of $f(x,y_0)$ while y_0 is by minimization of $f(x_0,y)$. This fact is the theoretical basis of the conventional procedure of the Lagrangean method where an optimum solution is sought as a stationary point of f(x,y) with respect to (x,y).

The point (x_0, y_0) which satisfies (A) is called *saddle point.*

The min-max theorem has a special meaning for the organization theory of this book. For instance, when IBP coordination principle is used and if the min-max theorem holds, the coordinability condition holds as shown in Chap.7.

AII.4 Min-max Theorem for Differentiable Function

If a function f(x,y) is sufficiently smooth (continuous) and if we can apply the differential calculus to it, the min-max theorem can be proven easily.

Suppose the following conditions hold:

(i) f is differentiable.

(ii) There exists $(x_0, y_0) \in X \times Y$ such that

$$\nabla_x f(x,y) = \nabla_y f(x,y) = 0 \text{ at } (x,y) = (x_0, y_0).$$

Then it is easy to prove that (x_0, y_0) is a saddle point.

Since f(x,y) is concave with respect to x for each y,

$$f(rx_1 + (1-r)x_2, y) \geqq rf(x_1, y) + (1-r)f(x_2, y)$$

holds for any $x_1, x_2 \in X$, r where $0 \leqq r \leqq 1$. That is, for $h = x_1 - x_2$

$$f(x_2 + h, y) - f(x_2, y) \leqq (f(x_2 + rh, y) - f(x_2, y))/r.$$

Then, taking the limit with respect to r we have

$$f(x_2 + h, y) - f(x_2, y) \leqq \nabla_x f(x_2, y)h.$$

Similarly, we have

$$f(x, y_2 + h) - f(x, y_2) \geqq \nabla_y f(x, y_2)h.$$

Then

$$f(x_0 + h, y_0) \leqq f(x_0, y_0) + \nabla_x f(x_0, y_0)h.$$

Since $\nabla_x f(x_0, y_0) = 0$,

$$f(x_0 + h, y_0) \leqq f(x_0, y_0)$$

for any h where $x_0+h \in X$.
Consequently,

$$\min_y \max_x f(x,y) \leqq \max_x f(x,y_0)=f(x_0,y_0).$$

Similarly,

$$\max_x \min_y f(x,y) \geqq \min_y f(x_0,y)=f(x_0,y_0).$$

Therefore,

$$\max_x \min_y f(x,y) \geqq f(x_0,y_0) \geqq \min_y \max_x f(x,y). \tag{B}$$

In general,

$$\max_x \min_y f(x,y) \leqq \min_y \max_x f(x,y). \tag{C}$$

as shown in Appendix I. Therefore, (B) and (C) imply

$$\max_x \min_y f(x,y) = \min_y \max_x f(x,y)$$

is true.

An elementary treatment of optimization is done on the domain of calculus. Although calculus is very powerful as shown above, its application is limited because it requires that functions which describe a model must be differentiable or an optimal point is hopefully an inner point rather than a boundary point. Many engineering systems satisfy the requirement but there are many cases in management problems where functions of a model cannot be differentiable or show singular properties with respect to differential operations. A typical example of the latter case is the LP (linear programming). Functions of an LP problem are linear which are differentiable but we cannot derive a solution of an LP problem by calculus. Since an optimal solution of an LP does not satisfy $\nabla_x f=\nabla_y f=0$ in general, a simple application of calculus is not effective to this situation.

AII.5 Min-max Theorem

There is another method to deal with optimization. It is an algebraic method which does not require the differentiability. It is more suitable for the organization theory than calculus although it requires more sophisticated mathematical facts than the latter. The conventional LP solution algorithm is based on the algebraic method.

This section will prove the min-max theorem in the algebraic way. We will proceed in the following steps: First we introduce the separating hyper plane theorem, which is quite simple conceptually but useful. Then, a technical lemma, Berge Lemma, will be proved. Finally, the min-max theorem will be proved using them.

Theorem AII.1 (Separating hyper plane theorem)

Let S_1 and S_2 be bounded closed convex sets in E_n. If $S_1 \cap S_2 = \phi$, there exists a hyper plane HP which separates S_1 and S_2, i.e.,

$$HP=\{x|px=r\}$$

and

$$(\forall x \in S_1)(px<r)$$

and

$$(\forall x \in S_2)(px>r)$$

where $p \in E_n$ and $r \in Re$.

Proof

Let

$$d(S_1,S_2)=\inf_{(x_1\in S_1,x_2\in S_2)} \| x_1-x_2 \| = \| x_0-y_0 \|$$

where $x_0 \in S_1$ and $y_0 \in S_2$. (Notice S_1 and S_2 are closed and bounded).

Since $S_1 \cap S_2 = \phi$, $\| x_0-y_0 \| >0$ or $x_0 \neq y_0$.

Let

$$HP=\{z|(y_0-x_0)(z-(y_0+x_0)/2)=0\}$$

that is, $p=(y_0-x_0)$ and $r=(y_0+x_0)(y_0-x_0)/2$.

First, we will show $HP \cap S_2 = \phi$. Suppose $z^*=HP \cap S_2$. Let $u=y_0-x_0$ and $v=z^*-x_0$. Then, since $z^* \in HP$,

$$(y_0-x_0)(z^*-(y_0+x_0)/2)=0,$$

or

$$(v+(v-u))u=0.$$

Consequently, we have

$$2uv=u^2.$$

Let

$$w=y_0+\lambda(z^*-y_0).$$

Then

$$\| w-x_0 \|^2 = \| u+\lambda(v-u) \|^2 = u^2+2u(v-u)\lambda+(v-u)^2\lambda^2 \equiv f(\lambda).$$

Let

$$a=(v-u)^2 \text{ and } b=u(v-u).$$

Then

$$f(\lambda)=a\lambda^2+2b\lambda+u^2.$$

Notice that $y_0 \neq z^*$. If $y_0=z^*$,

$$(y_0-x_0)(y_0-(y_0+x_0)/2)=0,$$

that is,

$$(y_0-x_0)(y_0-x_0)/2=0.$$

Then,

$$y_0=x_0$$

which contradicts $S_1 \cap S_2 = \phi$. Hence,

$$a=(v-u)^2=(z^*-y_0)^2>0.$$

Consequently, $f(\lambda)$ takes the minimum at

$$\lambda^* = -b/a = -u(v-u)/a$$

$$=(u^2-uv)/a$$

$$=(u^2-u^2/2)/a$$

$$=u^2/(2a).$$

Then, $\lambda^* \geqq 0$.

On the other hand,

$$1-\lambda^*=1+b/a$$

$$=((v-u)^2+u(v-u))/a$$

$$=(v^2-uv)/a$$

$$=(v^2-u^2/2)/a$$

$$=((v^2-u^2)+v^2)/(2a).$$

Since $z^* \in S_2$,

$$\| z^*-x_0 \| \geqq \| y_0-x_0 \| .$$

That is

$$v^2 \geqq u^2.$$

Hence,

$$1-\lambda^* \geqq 0.$$

Since $y_0, z^* \in S_2$ and $0 \leqq \lambda^* \leqq 1$,

$$w^* = y_0 + \lambda^*(z^* - y_0) \in S_2.$$

Since

$$\| w^* - x_0 \|^2 = f(\lambda^*)$$

$$= a(-b/a)^2 + 2b(-b/a) + u^2$$

$$= -b^2/a + u^2 < u^2,$$

(because $a > 0$ and $b = u(v-u) = uv - u^2 = u^2/2 - u^2 = -u^2/2 < 0$)
we have

$$f(\lambda^*) < u^2$$

which contradicts $u^2 = d(S_1, S_2)^2$. Therefore, $HP \cap S_2 = \phi$.
$HP \cap S_1 = \phi$ can be proven in the same way.
We will show that if $x \in S_2$,

$$(y_0 - x_0)(x - (y_0 + x_0)/2) = u(x - x_0) - u^2/2 > 0.$$

Since $S_2 \cap HP = \phi$,

$$u(x - x_0) - u^2/2 \neq 0.$$

Suppose $u(x - x_0) - u^2/2 < 0$. Let

$$q = y_0 + \lambda(x - y_0)$$

where $0 \leqq \lambda \leqq 1$. Then, since S_2 is a convex set, $q \in S_2$. Let

$$f(\lambda) = u(q - x_0) - u^2/2.$$

Then

$$f(0) = u^2 - u^2/2 > 0$$

and due to the assumption

$$f(1)=u(x-x_0)-u^2/2<0.$$

Since $f(\lambda)$ is a continuous function,

$$(\exists\lambda_0)(0<\lambda_0<1 \text{ and } f(\lambda_0)=0).$$

Then,

$$y_0+\lambda_0(x-y_0)\in HP$$

and hence

$$y_0+\lambda_0(x-y_0)\in HP\cap S_2$$

holds, which contradicts $S_2\cap HP=\phi$. Therefore, $(y_0-x_0)(x-(y_0+x_0)/2)>0$.

In the same way we can prove that if $x\in S_1$, $(y_0-x_0)(x-(y_0+x_0)/2)<0$.
Q.E.D.

We will introduce Berge's Lemma in a restricted form.

Lemma AII.1 [30]

Let M and K_i ($i=1,..,n$) be closed bounded convex sets in E_n. Suppose for any j, $1\leqq j\leqq n$,

$$M\cap[\cap^n_{(i=1,i\neq j)}K_i]\neq\phi$$

and

$$M\cap[\cap^n_{(i=1)}K_i]=\phi.$$

Then we have

$$M\not\subset[\cup^n_{(i=1)}K_i].$$

Proof

We will use the mathematical induction.

Let n=1. Then

$$M\cap K_1=\phi \text{ and } M\neq\phi$$

implies

$$M \not\subset K_1.$$

Suppose the lemma is true for $i \leqq n$. Let for any $j, 1 \leqq j \leqq n+1$,

$$M \cap [\cap_{(i=1, i \neq j)}^{n+1} K_i] \neq \phi$$

and

$$M \cap [\cap_{(i=1)}^{n+1} K_i] = \phi.$$

Then, since

$$(M \cap K_{n+1}) \cap [\cap_{i=1}^{n} K_i] = \phi$$

and since $(M \cap K_{n+1})$ and $\cap_{i=1}^{n} K_i$ are closed, bounded and convex, we have a separating hyper plane HP for them by the separating hyper plane theorem.

Let

$$M' = M \cap HP$$

and

$$K_i' = K_i \cap HP.$$

Then

$$M' \cap [\cap_{i=1}^{n} K_i'] = M \cap HP \cap [\cap_{i=1}^{n} (K_i \cap HP)]$$

$$M \cap (HP \cap [\cap_{i=1}^{n} K_i] = \phi$$

holds.

Since $M \cap [\cap_{i=1}^{n} K_i] \neq \phi$ due to the induction hypothesis, there is $y \in M \cap [\cap_{i=1}^{n} K_i]$.

Let $j \neq n+1$ be arbitrary. Since $M \cap [\cap_{(i=1, i \neq j)}^{n+1} K_i] \neq \phi$, there is

$$x_j \in M \cap [\cap_{(i=1, i \neq j)}^{n+1} K_i].$$

Since $M \cap [\cap^n_{(i=1,i\neq j)} K_i]$ is convex and $y, x_j \in M \cap [\cap^n_{(i=1,i\neq j)} K_i]$,

$$z = \lambda x_j + (1-\lambda) y \in M \cap [\cap^n_{(i=1,i\neq j)} K_i]$$

holds for $0 \leqq \lambda \leqq 1$. Since $x_j \in M \cap K_{n+1}$ and $y \in [\cap^n_{i=1} K_i]$, HP separates y and x_j, and hence there is $\lambda_j, 0 \leqq \lambda_j \leqq 1$ such that

$$z_j = \lambda_j x_j + (1-\lambda_j) y \in HP$$

and

$$z_j \in M' \cap [\cap^n_{(i=1,i\neq j)} K_i'].$$

Consequently, the induction hypothesis implies

$$M' \not\subset \cap^n_{i=1} K_i'$$

or

$$HP \cap M \not\subset HP \cap [\cup^n_{i=1} K_i].$$

Since $HP \cap M \cap K_{n+1} = \phi$,

$$M \cap HP \not\subset HP \cap [\cup^{n+1}_{i=1} K_i]$$

or we have

$$M \not\subset [\cup^{n+1}_{i=1} K_i].$$

Q.E.D.

Finally we will prove the min-max theorem.

Theorem AII.2 (Min-max theorem) [30]
Let X and Y be convex closed bounded subsets of linear spaces. Let $f: X \times Y \rightarrow Re$ be a continuous function on $X \times Y$. Suppose f is a convex with respect to x for each y and a concave with respect to y for each x. Then, the following hold:

$$max_y\, min_x f(x,y) = min_x\, max_y f(x,y).$$

Proof

Suppose the conclusion is false, that is there is $c \in Re$ such that

$$\max_y \min_x f(x,y) < c < \min_x \max_y f(x,y).$$

Let

$$Y_x = \{y | f(x,y) < c\}.$$

Since $\max_y \min_x f(x,y) < c$, we have

$$Y \subset \cap_x Y_x .$$

Since Y is closed and bounded, there exists $X_l = \{x_1,..,x_n\}$ such that

$$\max_Y \min_{X_l} f(x,y) < c .$$

Similarly, there exists $Y_l = \{y_1,..,y_m\}$ such that

$$\min_X \max_{Y_l} f(x,y) > c$$

holds.

Since $X_l^h \subset X$ and $Y_l^h \subset Y$, we have

$$\max_{Y_l^h} \min_{X_l} f(x,y) < c < \min_{X_l^h} \max_{Y_l} f(x,y) .$$

Let $Y_2 \subset Y_l$ be a minimal such that

$$c < \min_{X_l^h} \max_{Y_2} f(x,y)$$

holds. Since Y_l is a finite, Y_2 always exists. Similarly, let $X_2 \subset X_l$ be a minimal such that

$$c > \max_{Y_2^h} \min_{X_2} f(x,y) .$$

In this way we can generate sequences $\{X_i\}$ and $\{Y_j\}$. Since, X_i and Y_i are finite, there is a limit (X_0, Y_0) such that

$$\max_{Y_0^h} \min_{X_0} f(x,y) < c < \min_{X_0^h} \max_{Y_0} f(x,y)$$

where $X_0 = \{u_1,..,u_p\}$ and $Y_0 = \{v_1,..,v_q\}$. Let

$$X_j = \{x | f(x,v_j) \leqq c, x \in X_0^h\}.$$

Then, since f(x,y) is convex and continuous and X_0^h is bounded, X_j is a closed, bounded and convex set. Furthermore, $c < \min_{X_0^h} \max_{Y_0} f(x,y)$ implies

$$X_0^h \cap [\cap_{j=1}^q X_j] = \phi$$

and since Y_0 is minimal,

$$X_0^h \cap [\cap_{(j=1,j\neq i)}^q X_j] \neq \phi$$

for any $1 \leqq i \leqq p$. As a matter of fact, if

$$X_0^h \cap [\cap_{(j=1,j\neq i)}^q X_j] = \phi$$

for some i, then $Y_0' = Y_0 - \{i\}$ satisfies

$$c < \min_{X_0^h} \max_{Y_0} f(x,y)$$

which contradicts the minimality of Y_0. Then, Lem. AII.1 implies that

$$X_0^h \not\subset \cup_j X_j$$

or

$$(\exists x_0 \in X_0^h)(\forall v_j \in Y_0)(f(x_0,v_j) > c).$$

Since f(x,y) is convex with respect y for each x, we have

$$(\exists x_0 \in X_0^h)(\forall y \in Y_0^h)(f(x_0,y) > c).$$

In the similar way we have

$$(\exists y_0 \in Y_0{}^h)(\forall x \in X_0{}^h)(f(x,y_0)<c).$$

Then,

$$f(x_0,y_0)< c <f(x_0,y_0)$$

holds, which is a contradiction.
Q.E.D.

APPENDIX III MAXIMUM PRINCIPLE

AIII.1 Introduction

Pontryagin's maximum principle is one of the most powerful tools for dynamic optimization when a model is described in an analytical way. Since an organization is a dynamic system, it needs a good optimization tool for its management. But because an organization is a poorly structured system which is not easily described by, for instance, a differential equation, a dynamic optimization by an analytical method cannot be directly applied to an organization. But since the maximal principle shows a basic framework how to find a solution of a dynamic optimization problem, it will be introduced in this appendix. This method is used in Chap. 5.

This appendix will introduce the principle in the most simple but maybe the most useful form [31].

AIII.2 Maximum Principle

Suppose the state equation of a dynamic system is given by

$$dz/dt=f(z,m)$$

where z is the state vector and m is the manipulating variable and $z(0)=z_0$ is fixed.

Suppose the objective function is given by

$$J=\int_0^T g(z,m)dt \rightarrow \max .$$

If we rewrite the state equation as

$$f(z,m)-dz/dt=0,$$

it can be considered as an equality constraint condition of the maximization. Then, let us apply the Lagrangean method discussed in Appendix II to the current optimization problem.

Let

$$L(z,\dot{z},m,p)=g(z,m)-p(f(z,m)-\dot{z})$$

and

$$H(z,m,p)=g(z,m)-pf(z,m)$$

where p is a Lagrangean multiplier which is a vector valued time function. pf(z,m) is an inner product of p(t) and f(z,m)(t). Then, the Lagrangean formulation is given by

$$\phi(z,\dot{z},m,p)=\int_0^T L(z,\dot{z},m,p)dt$$

where

$$\int_0^T P(f(z,m)-\dot{z})dt$$

is the Lagrangean multiplier term. Since $\dot{z}$ is uniquely determined by z, the Lagrangean function had better be described as ϕ (z,m,p) rather than ϕ (z, $\dot{z}$,m,p) but we follow the convention.

Now we apply the variational calculus to ϕ (z, $\dot{z}$,m,p) on function spaces. Let

$$\delta\, z=sh,$$

$$\delta\, m=s_2h_2$$

and

$\delta p = s_3 h_3$

where s, s_2, s_3 are small real numbers while h, h_2, h_3 are arbitrary smooth(differentiable) time functions.

Then, since

$$L(z + sh, \dot{z} + s\dot{h}, m + s_2 h_2, p + s_3 h_3) =$$
$$L(z, \dot{z}, m, p) + L_z sh + L_{\dot{z}} s\dot{h} + L_m s_2 h_2 + L_p s_3 h_3 + o(h^2, h_2^2, h_3^2),$$

the first variation of J is given by

$$\delta J = \int_0^T (L_z sh + L_{\dot{z}} s\dot{h} + L_m \delta m + L_p \delta p) dt .$$

When the partial integration is applied to the second term of the right hand side of the first variation, we have

$$\int_0^T L_{\dot{z}} s\dot{h} dt = L_{\dot{z}} sh|_0^T - \int_0^T (dL_{\dot{z}}/dt) sh dt .$$

Consequently,

$$\delta J = \int_0^T ((L_z - (dL_{\dot{z}}/dt)) sh + L_m \delta m + L_p \delta p) dt + L_{\dot{z}} sh|_0^T .$$

Since h, h_2, h_3 are arbitrary, the condition that J takes an extremum at z, m, p requires that

$$L_z - (dL_{\dot{z}}/dt) = 0 ,$$

$$L_m = 0 ,$$

$$L_p = 0,$$

and

$$L_{\dot{z}} sh |_0^T = 0 .$$

Since $L=H(z,m,p)-p\dot{z}$, the above conditions imply

$$\dot{p}=-H_z\,,$$

$$\dot{z}=f(z,m),$$

and

H is maximized with respect to m, respectively.
Furthermore, since $L_{\dot{z}}sh\,|_0^T=0$ implies

$$-ph\,|_0^T=0$$

and since h(0)=0 (because z(0) is fixed), and h(T) is arbitrary,

$$p(T)=0$$

should hold.

In summary, a necessary condition of finding an optimum solution of the dynamic problem is given by finding (z,m,p) which satisfies the following differential equations

$$\dot{p}=-H_z$$

and

$$\dot{z}=f(z,m)$$

where m is specified by maximization of H for each given (z,p) and the boundary conditions are

$$z(0)=z_0$$

and

$$p(T)=0.$$

References

1. K. M. Carley, "Computational and Mathematical Organization Theory:Perspective and Directions", Computational and Mathematical Organization Theory, No.1, Vol.1, 1995
2. M. D. Mesarovic and Y. Takahara, Abstract Systems Theory, Lecture Notes in Control and Information Science, Springer, 1989
3. F. E. Kast and J. E. Rosenzweig, Organization and Management:A Systems and Contingency Approach, McGraw-Hill, 1981
4. J. G. Miller, Living Systems Theory, McGrawHill, 1970
5. S. Beer, Diagnosing the system for organizations, John Wiley, 1985
6. R. Espejo et. al, Organizational Transformation and Learning, Wiley, 1996
7. W. R. Ashby, Introduction to Cybernetics, Methuen, 1964
8. M. Mesarovic, D. Macko and Y. Takahara, Theory of Hierarchical Multilevel Systems, Academic Press, 1974
9. J. Bridge, Beginning Model Theory, Oxford University Press, 1978
10.M. D. Cohen, J. B. March and J. P. Olsen, "A Garbage Can Model of Organizational Choice", Administrative Science Quarterly, 17(1)
11.Pierre N. V. Tu, Dynamical Systems, Springer, 1994
12.H. Simon, "Administrative Behavior", Free Press, 1957
13.L. J. Savage, The Foundation of Statistics, John Wiley,1954
14.M. J. Beckman, Tinbergen Lectures on Organization Theory, Springer, 1988
14.R. Repullo, "A Simple Proof of Maskin's Theorem on Nash Implementation", Social Choice and Welfare, 4, 1987
16.R. F. Drenick, A Mathematical Organization Theory, North-Holland, 1986
17.K. Kijima,Y. Takahara and B. Nakano, "Algebraic Formulation of Relationship Between a Goal Seeking System and Its Environment", Int. J. of General Systems, Vol.12, 1986
18.K. C. Laudon and J. P. Laudon, "Management Information Systems", Prentice Hall, 2000

19.L. Bertallanffy, General Systems Theory, Braziller, 1968
20.J. W. Forrester, Principles of Systems, Wright-Allen, 1969
21.J. P. Checkland, Systems Thinking, Systems Practice, John Wiley, 1981
22.K. E. Boulding, "General systems theory-the skeleton of science", Management Science, 2(3), 1956
23.E. Lazzlo, Introduction to Systems Philosophy, Braziller, 1972
24.L. A. Zadeh, "Fuzzy Sets", Information and Control, 8, 1965
25.G. J. Klir, An Approach to General Systems Theory, Van Nostrand, 1969
26.W. Wymore, Model-Based Systems Engineering, CRC Press, 1992
27.F. Pichler and H. Schwaertzel, CAST Methods in Modeling, Springer, 1992
28.X. Wu, The Pansystems View of the World, People University Press, 1990
29.S. Takahashi and Y. Takahara, Logical Approach to Systems Theory, Lecture Notes in Control and Information Science, Springer, 1995
30.J. Stoer and C. Wizgall, Convexity and Optimization, Springer, 1970
31.Y. Takahara, N. Shiba and Y. Liu, "Mathematical General Systems Theory and Information Systems", Internal report, Dep. of Industrial and Systems Eng., Chiba Institute of Technology, Chiba,Japan,2000
32.Y. Takahara, N. Shiba and Y. Liu,"extProlog for a Problem Solving EUD Methodology", Internal report, Dep. of Industrial and Systems Eng., Chiba Institute of Technology, Chiba, Japan,2000
33.Y. Takahara, X. Chen and N. Shiba, Advances of DSS Development-General Systems Approach, Economics and Science Publisher, Beijin, 2003

Index